Industry 4.0 Technologies for Business Excellence

Demystifying Technologies for Computational Excellence: Moving Towards Society 5.0

Series Editors:
Vikram Bali and Vishal Bhatnagar

This series encompasses research work in the fields of Data Science, Edge Computing, Deep Learning, Distributed Ledger Technology, Extended Reality, Quantum Computing, Artificial Intelligence, and various other related areas such as natural-language processing and technologies, high-level computer vision, cognitive robotics, automated reasoning, multivalent systems, symbolic learning theories and practice, knowledge representation and the semantic web, intelligent tutoring systems and education.

The prime reason for developing and growing out this new book series is to focus on the latest technological advancements – their impact on society, the challenges faced in implementation, and the drawbacks or reverse impact on society due to technological innovations. With these technological advancements, every individual has personalized access to all services and all devices connected with each other communicating amongst themselves, thanks to the technology for making our life simpler and easier. These aspects will help us to overcome the drawbacks of the existing systems and help in building new systems with the latest technologies that will help society in various ways, proving Society 5.0 as one of the biggest revolutions in this era.

Computing Technologies and Applications
Paving Path Toward Society 5.0
Edited by Latesh Malik, Sandhya Arora, Urmila Shrawankar, Maya Ingle, and Indu Bhagat

Reinvention of Health Applications with IoT
Challenges and Solutions
Edited by Dr. Ambikapathy, Dr. Shobana, Dr. Logavani, and Dr. Dharmasa

Healthcare and Knowledge Management for Society 5.0
Trends, Issues, and Innovations
Edited by Vineet Kansal, Raju Ranjan, Sapna Sinha, Rajdev Tiwari, and Nilmini Wickramasinghe

For more information on this series, please visit: https://www.routledge.com/Demystifying-Technologies-for-Computational-Excellence-Moving-Towards-Society-5.0/book-series/CRCDTCEMTS

Industry 4.0 Technologies for Business Excellence
Frameworks, Practices, and Applications

Edited by
Shivani Bali
Sugandha Aggarwal
Sunil Sharma

CRC Press
Taylor & Francis Group
Boca Raton London New York

CRC Press is an imprint of the
Taylor & Francis Group, an **informa** business

First edition published 2022
by CRC Press
2 Park Square, Milton Park, Abingdon, Oxon, OX14 4RN

and by CRC Press
6000 Broken Sound Parkway NW, Suite 300, Boca Raton, FL 33487-2742

© 2022 selection and editorial matter, Shivani Bali, Sugandha Aggarwal, Sunil Sharma; individual chapters, the contributors

CRC Press is an imprint of Informa UK Limited

The right of Shivani Bali, Sugandha Aggarwal, Sunil Sharma to be identified as the author[/s] of the editorial material, and of the authors for their individual chapters, has been asserted in accordance with sections 77 and 78 of the Copyright, Designs and Patents Act 1988.

All rights reserved. No part of this book may be reprinted or reproduced or utilized in any form or by any electronic, mechanical, or other means, now known or hereafter invented, including photocopying and recording, or in any information storage or retrieval system, without permission in writing from the publishers.

For permission to photocopy or use material electronically from this work, access www.copyright.com or contact the Copyright Clearance Center, Inc. (CCC), 222 Rosewood Drive, Danvers, MA 01923, 978-750-8400. For works that are not available on CCC, please contact mpkbookspermissions@tandf.co.uk

Trademark notice: Product or corporate names may be trademarks or registered trademarks and are used only for identification and explanation without intent to infringe.

ISBN: 978-0-367-69117-2 (hbk)
ISBN: 978-0-367-69118-9 (pbk)
ISBN: 978-1-003-14047-4 (ebk)

DOI: 10.1201/9781003140474

Typeset in Times
by codeMantra

Contents

Preface vii
Editors xv
Contributors xvii

Chapter 1 Understanding the Industry 4.0 Revolution Using Twitter Analytics ... 1

Jatinder Bedi, R. K. Padhy, and Sidhartha S. Padhi

Chapter 2 The Role of Universal Product Coding (UPC), Global Data Synchronization Network (GDSN) and Product Category Management in Efficient Consumer Response (ECR) 27

Sunil Sharma

Chapter 3 Delivering Superior Customer Experience through New-Age Technologies 47

Vaishali Kaushal and Rajan Yadav

Chapter 4 Use of Artificial Intelligence-Enabled Features in the Retail Sector: A Perceptual Study of Customers 61

Ashutosh Mohan, Upnishad Mishra, and Ishi Mohan

Chapter 5 Effective Integration of Lean Operations and Industry 4.0: A Conceptual Overview 97

Aaron Ratcliffe, Maneesh Kumar, and Sriram Narayanan

Chapter 6 Opportunities and Risks: Use of Autonomous Vehicles in Logistics 115

Nikunj S. Yagnik

Chapter 7 Assessment of Challenges for Implementation of Industrial Internet of Things in Industry 4.0 127

Snigdha Malhotra, Tilottama Singh, and Vernika Agarwal

Chapter 8 IoT Security Issues and Solutions with Blockchain 141

Arvind Panwar, Vishal Bhatnagar, Sapna Sinha, and Raju Ranjan

Chapter 9 Stabilization of Imbalance between the Naira and the Dollar Using Game Theory and Machine Learning Techniques 163

Garba Aliyu, Bukhari Badamasi, Sandip Rakshit, and Onawola, H.J.

Chapter 10 The Emerging Role of Big Data in Financial Services 175

Deepika Dhingra and Shruti Ashok

Chapter 11 Digital Payments in India: Impact of Emerging Technologies 191

Manisha Sharma

Chapter 12 Cryptocurrency: Perspectives, Applications, and Issues 205

Abhishek Sharma, Ayush Srivastava, and Deepika Dhingra

Chapter 13 Models for Predicting Student Enrolment for Delhi-Based Schools 221

Kartik Kakani, Biswarup Choudhury, and Sugandha Aggarwal

Chapter 14 Analyzing the Functionality and Efficient Operability of the Youth During COVID 19 237

Megha Mishra, Reema Thareja, and Vidushi Singla

Chapter 15 AI in Talent Management for Business Excellence 255

Subhajit Bhattacharya

Index 267

Preface

Greetings!

The fourth Industrial Revolution, Industry 4.0, created by Big Data and emerging technologies, has altered the way businesses function. Organisations now thrive by using the valuable insights gained from the vast consumer data. In the present data-driven world, analytics play a vital role in accomplishing business targets by turning user data into key learnings and formulating strategies to make smart decisions in business. There is a huge investment being made all over the world by companies in the analytics area and its applications to leverage their power to gain business profitability.

Industry 4.0 is driven by a simple idea: merging an existing physical entity with advanced IT elements such as the Internet of Things (IoT), Cloud-Based Data, Sensors, Automators, and Analytics; creating something that offers a surplus benefit over its original predecessor, Industry 3.0. Machine Learning, Cloud Computing, Cyber Security and IoT are the topics of contemporary research and industry application interest.

This book addresses the key issues and themes related to applications of Industry 4.0 in the domain of Marketing, Operations & Supply Chain, Finance and HR for achieving business excellence spread over 15 chapters. It brings together researchers, developers, practitioners and users who are interested in these areas to explore new ideas, techniques, and tools and to exchange their experiences. The primary aim is to convey the ideas that have emerged recently in Industry 4.0 applications in the field of business excellence.

This edition results from extensive revisions throughout the text. The material is received from eminent authors and practitioners working in the field of Industry 4.0. The research on Industry 4.0 is expanding so rapidly now that the volume of potential new research papers is vast. However, the relevance and application of research papers have been kept as a focus while finalising the contributions from a wide spectrum of authors to this volume.

Chapter 1, "Understanding the Industry 4.0 Revolution using Twitter Analytics", focuses on Twitter as one of the social networking sites to improve understanding of the fourth industrial revolution and its associated initiatives in the context of social media. These tweets were collected using the hashtags #Industry4.0, #factoroffuture, #AdvanceManufacturing. This is followed by Twitter Analytics (TA) that comprises three types of analytics, namely descriptive statistics, content analytics that includes sentiment analytics and emotional score analysis, and network analysis and topic modelling. This methodology is applied to around 31000 tweets to answer four research questions: (1) Which tweets characterize Industry 4.0? The finding suggests that a great part (35%) of #Industry4.0 tweets contains more than one hashtag. Primarily, tweets about the convenient issues and difficulties with Industry 4.0 applications are high in numbers. (2) What are the topics shared on Twitter? Findings suggest that most mainstream hashtags are like themed territories canvassed in scholarly diaries. They incorporate #IOT, #IIOT, #Bigdata, #Analytics, #Cybersecurity, #Innovation,

#Automation, #digital, #sensors, #cyberattacks, #smart city, #Block chain, #5G, #drone, #robotics, #RPA, #ML, #deep learning, #cloud and #AI. Few noticeable ones were # PlanetEarthFirst #SustainableDevelopment and #Futureofwork. (3) What are the features of the users who use Industry 4.0 related tweets? Findings suggest that some of the clients are unemployed while others are professionals. Almost all the tweets under this group of clients are unique and are experts in innovation-related jobs. Lastly (4) What are the sentiments of these tweets? Findings suggest that the tweets data test for #Industry4.0 consists of moderately positive feeling emphasizing about occasions, fabricating capacities news, business use case, changes, reports, innovations, abilities, and administrations notices. Finally, the future scope and limitations of the study are also reported.

In Chapter 2, "Role of Universal Product Coding (UPC), Global Data Synchronisation Network (GDSN) and Product Category Management in Efficient Consumer Response (ECR)", it is elaborated that the ECR – Efficient Consumer Response, is a shared initiative among manufacturing and service companies and institutions, to respond to the consumers' needs at a short notice and to reduce the unnecessary costs and long cycle times. The ECR aims to accomplish real-time collaboration amongst supply chain members needed to satisfy consumers' needs faster, better and cheaper. The paper first analyzes ECR, its various components and prerequisite activities. The specific role of product category management, especially in fast-moving consumer goods, is being studied. In the current digitalised environment, the role of data flow across supply chain partners through global data synchronisation network (GDSN) with universal product coding (UPC) as an enabler is also highlighted. Finally, the Indian and global scenario in ECR is also mapped.

In Chapter 3, "Delivering superior Customer Experiences through new age technologies", it is elaborated that the unique power of modern technologies and virtual consumers are indicators of a new marketing paradigm. Customer Experience (CX) is at the heart of a successful product or service offering. To deliver an enhanced experience, today, brands have incorporated several smart technologies like radio frequency identification systems (RFID), smart shopping carts, Augmented Reality (AR), and virtual reality, among others. Augmented reality's incorporation of immersive real-time virtual content into the consumer's view of the physical world makes for a superior customer experience. Artificial Intelligence-based technologies help consumers better understand their own expectations. The AI-empowered and Natural Language processing stimulated Chatbots are designed to comprehend human language and address the consumer's appetite for immediate responses to queries. The article confers how these new age technologies are transforming the customer experience. In addition, the article attests to industry examples who are benefitting from these new age technologies like AR, AI and Chatbots. It will also explain the challenges faced by organizations during the digital journey by employing multiple new-age technologies adopted in 2021.

In Chapter 4, "Use of Artificial Intelligence Enabled Features in the Retail Sector: A Perceptual Study of Customers", it is explained that marketers have widely accepted the use of Artificial Intelligence (AI) in marketing. Despite an increase in the use of AI-based tools and features by online retailers, knowledge of these technologies' customer perspective is far from conclusive. Existing research in the field tends to focus mainly on the technical aspects of AI. However, there is little evidence that the researchers have tried to explore the customer perspective towards AI. The chapter is aimed at exploring customer perceptions of AI. Different features have been identified during the study that customers find significant for them, such as voice search, personalized marketing, content automation, etc. The chapter tries to find customers' perceptions of these AI-enabled features and determine customer satisfaction. The possible results may help the retailers identify and integrate those tools into their platform, which the customer finds useful and improves the one in which they are least interested.

In Chapter 5, "Effective Integration of Lean Operations and Industry 4.0: A Conceptual Overview", it is elaborated that lean Implementation is an important recognized aspect of supply chain management. Specifically, Lean principles offer a means for maintaining focus on operational effectiveness in the era of rapid technological advancement accelerated by Industry 4.0. New Industry 4.0 capabilities offer a means for enhancing the strength and expanding the scope of Lean practices through value chain integration, automation, and digitization. Lean helps improve productivity and other performance metrics by reducing waste existing in business processes and its supply chain using bundles of soft and hard practices including just-in-time manufacturing, set-up, pull, flow, total productive maintenance, long-term strategic relationships with suppliers and customers, and statistical process control. By applying both Lean and Industry 4.0 together, organizations can find ways to increase productivity and stretch the boundaries of the efficient frontier. This chapter discusses how to employ Lean principles and augment Lean practices by adopting Industry 4.0 technologies. Specifically, overlaps are found between Lean Principles and Industry 4.0 and showcase real-world examples of how the two are deeply intertwined.

In Chapter 6, "Opportunities & Risks: Use of Autonomous Vehicles in Logistics", it is elaborated that the crux of the autonomous vehicles to be part of logistics is giving intelligence to various segments of logistics, mainly objects to select the path for themselves to follow in a logistics network according to the tasks set by the organization, and can accomplish decisions based on local information and in real time. This paper gives a portrayal of the use of autonomous vehicle as a new tool to adapt to the complexity and dynamics of recent logistics systems in terms of risks and opportunities associated with them. While utilizing such autonomous vehicles in various segments of logistics leads to generic and specific risks which are delineated and exemplarily scrutinized. However, utilization of such innovations may impact prevailing risks or introduce unknown dangers into the coordination framework too. This paper solely gives information about methodology, material, utilization and outcomes while utilizing intelligent vehicles in logistics.

In Chapter 7, "Assessment of Challenges for Implementation of Industrial Internet of Things in Industry 4.0", it is elaborated that advancement in digital technologies has presented a new outlook to integrate optimization of operational efficiency and automation into supply chain operations. Industrial Internet of Things (IIoT) can aid in bringing together brilliant machines, advanced analytics, and the people involved in work. The goal of industrial digital transformation is to serve as a new vision of Internet of Things by combining machine-to-machine communication with big data analytics in an industry. It can drive unprecedented levels of efficiency, productivity, and performance. However, there is a substantial gap when it comes to examining the implantation and results of IIOT techniques in Indian industries, as studies have rightly demonstrated that prior changes, planning and re-designing of the system are essential to ensure any change, aiming to reduce the failure rate. In this chapter, the aim is to comprehend the challenges faced by Indian enterprises in India in implementing IIoT in their business operations. The methodology of grey Decision-making trial and evaluation laboratory (DEMATEL) is employed to understand the interrelationship between these challenges and to identify them as cause and effect. The present study is validated for enterprises across the national capital region. India.

In Chapter 8, "IoT Security Issues and solutions with blockchain", it is elaborated that in today's world, our life is simplified by the diverse availability of various gadgets or equipment. Linking all these gadgets through the Internet and ultimately commanding them remotely through smartphone apps or other similar means will make our lives more comfortable and simpler. With the arrival of smart cities, smart power grids, smart homes, and whole smart things around us, the IoT has arisen as a field of unbelievable potential, impact, and exponential growth. Although a maximum of these devices is comfortable to compromise security and hack. Classically, all the IoT devices have less computing power, less storage space, and limited network capacity, which means they are very vulnerable to hacking. IoT devices are easy to hack compared to endpoint devices like computers, laptops, smartphones, etc. In this paper, the author presents a survey on security issues in IoT. The author reviews security issues and categorizes them with IoT reference architecture. Most importantly, the author presents how blockchain can play a vital role in resolving various IoT security issues. The paper also classifies research challenges and security concerns related to security in IoT.

In Chapter 9, "Stabilization of Imbalance between Naira and Dollar Using Games Theory and Machine Learning Techniques", it is elaborated that the instability of the Naira against foreign currencies has been a major setback that has sabotaged the revival process of stabilizing the country's economy. This is coming because of the high importation of goods that necessitated an exchange of Naira with other foreign currencies, particularly the dollar. The Nigerian economy solely relies on the exportation of crude oil. However, Nigeria imports refined petroleum products for internal consumption, which requires the use of billions of dollars. Moreover, there is no strike balance between what we import from and export to foreign countries. The stabilization of Nigeria will help strengthen the country's economy by making the right policy decisions. Such a decision encompasses the domestic refining of crude oil and exploration of other means such as the Industrial Revolution, Agricultural sector, Information Technology & Computing, among others. Because of the fluctuation of Naira, foreign investors are scared of coming to invest. Moreover, some citizens would rather invest outside Nigeria than invest locally. This paper proposes a conceptual model using game theory and machine learning to address the aforementioned challenges. The game theory is used to resolve the conflict between the Naira and other foreign currencies for the stability of Naira and machine learning to build a model that will predict the possibility of appreciation or depreciation of Naira based on the government's policy put in place.

In Chapter 10, "Emerging Role of Big Data in Financial Services", it is explained that the financial services industry has always been a data-intensive industry, generating large quantities of customer data from millions of transactions conducted every day. This vast explosion of data and the surge in technological complexities is transforming the way the financial sector operates and thus big data management of various financial products and services is emerging as a potential field of study. The application of big data in financial services has abundant advantages for financial institutions: improved customer engagement, better fraud detection and enhanced market trading analysis. This chapter presents applications of big data, stimulating various financial sectors, highlighting its influence on financial markets, financial institutions, fraud detection, risk analyzis, etc. The association between big data and finance-related mechanisms is analyzed through an exploratory literature assessment using secondary data. This chapter also describes real world, contemporary case studies on the applications of Big Data in the financial world. The findings demonstrate that there are many technological research challenges at all levels of the big data chain that need to be developed to provide more competitive and effective solutions. Since big data is a relatively new concept in the field of finance, this study also provides directions for future research.

In Chapter 11, "Digital Payments in India: Impact of Emerging Technologies", it is elaborated that the world has changed the way it worked primarily due to the technological revolution. Information and Communication Technologies, along with the rise of the Internet, have opened new avenues of business, and one of the pathbreaking innovations to date is the mode of transaction using the technology, and that is digital payments. Currency exchange has witnessed the novelty of ideas in the digital sector over the past few years, but contactless digital payment technology has changed the dynamics of the banking industry. The evolution of these payment methods is meant to transform the traditional digital wallet. The article therefore explores the inhibitors and facilitators of digital payments which may eventually be major push or pull factors in drawing users of digital payments in the long run. The article further discusses the impacts of integration of digital payments with emerging technologies such as Blockchain, big data analytics, social media analytics and cloud computing that might contribute to the growth of digital payments in India. The discussions and the observations will be useful to the Indian government's initiative of Digital India and will provide substantial inputs to government agencies, financial institutions, mobile & telecommunications operators and researchers.

In Chapter 12, "Cryptocurrency: Perspectives, Applications, and Issues", it is elaborated that the recent development of communication and information technologies has transformed the lives of people. The Internet has supported people in various ways to become more flexible and effective for carrying out tasks, from booking a cab to transferring huge amounts of money in just a few clicks. A surge in the users of the internet has led to the inception of notions and devised a new exchange dimension of cryptocurrency. This paper provides diverse perspectives on cryptocurrency and its adoption. It also aims to scrutinize the terms of regulation and legislation, issues faced, and various scams involving cryptocurrencies. It also discourses on the potential & implications of cryptocurrency.

In Chapter 13, "Models for Predicting Student Enrolment for Delhi Based Schools", it is elaborated that the enrolment rate in schools within our country is in decline, and the authors further recommend how the enrolment rate can be increased in schools through the improvement of basic school facilities and features such as furniture available for students, number of classrooms available for students, ICT Lab, laptops, and Digi board. For the analysis of the same, data has been collected from a government website that contains information about various schools in and around Delhi. Also in this chapter, multiple regression techniques such as Linear Regression, Ridge Regression, Lasso Regression ElasticNet Regression were used on the data that was collected to predict the school enrolment, and in the end, comparison of different regression techniques took place to determine the best predicting model. Also, in the end, certain recommendations are given to the Government about how they can increase school enrolment rates by giving certain features more importance than others.

In Chapter 14, "Analyzing the Functionality and Efficient Operability of the Youth During Covid 19", it is elaborated that the ongoing pandemic has increased stress and problems for people of all age groups. There has been a drastic change in the life of people due to this pandemic because of which they are suffering from mental stress, financial problems, anxiety, etc. Different measures such as lockdown and social distancing were adopted by almost every country to control the pandemic. People have lost jobs, have had salary cuts, and are suffering great financial losses due to these measures, which are causing mental stress to people. The youth are also facing many problems due to the closing of educational institutions, a sudden shift towards an online mode of education, and the non-availability of jobs due to the falling economy in COVID. Hence, a questionnaire was prepared about the problems faced by people. The questions were divided into three sections: education, health and lifestyle. The questionnaire was filled by individuals in the age group 17-30 in India. Responses were collected and analyzed using statistical techniques and machine learning algorithms such as KNN, Logistic Regression, SVM. To compare the situations of the people in India with those in the other parts of the globe, data was extracted from Twitter. Results were plotted graphically to get a better understanding of the data and the result set. Accuracy was calculated and a confusion matrix was drawn to validate our calculations and conclusions.

In Chapter 15, "AI in Talent Management for Business Excellence", it is elaborated that today with the advent of technologies and global business competitiveness, every organization is trying to bring its momentum at a competitive pace to ensure long-term sustainability, client retention, and global presence. However, it is not only the technology that is playing a crucial role but also the committed and self-motivated workforces who are playing pivotal roles to ensure the multifaceted growth of an organization. In an organization, workforces are augmented with talents, thus it's imperative to procure, nurture, cultivate, and harvest them efficiently. Leading organizations are rigorously working on the next generation AI-led solutions for end-to-end talent management to perform workforce orchestration and optimization in real-time. The intelligent system is so designed that it can integrate and synchronize data from disparate platforms. The system while working on a hybrid SaaS model is highly scalable and agile in nature and can be customized as per the organizational needs. Intelligent AI Agents at the client end responsively get integrated with the client data and sets a bridge between the client and server for data exchange while empowering advanced AI-led analytics services. This intelligent application helps the workforce mobilized, *dynamized*, and *optimized* holistically, called as Intelligent Workforce Orchestrator and Advisor (IWOA).

Hope, this volume goes a long way in adding some new dimensions to the framework, practices and applications of Industry 4.0 for achieving business excellence. We thank Taylor and Francis to take up this project with us and bringing it in your hands in short time. We thank our respective spouses and families for their support in enabling us to bring about this volume.

Dr. Shivani Bali,
Dr. Sugandha Aggarwal,
Dr. Sunil Sharma

Editors

Dr. Shivani Bali is working as a Professor in Business Analytics area at Jaipuria Institute of Management, Noida, India. She is an academician, trainer, researcher, and consultant with more than 17 years of experience in the fields of Business Analytics and Operations Management. She has received her Bachelor's, Master's, and Ph.D. degrees from the University of Delhi, India. She has published around 30 research papers indexed in SCI/WoS/Scopus journals and published 4 patents. She has been working on consulting assignments with many reputed companies and has imparted training programs in the areas of data science and business analytics. Her areas of interest are Business Analytics, Data Science, Operations, and Supply Chain Analytics.

Dr. Sugandha Aggarwal is an Assistant Professor in Operations and Analytics Department and Programme Chair of PGDM (Research and Business Analytics) at Lal Bahadur Shastri Institute of Management (LBSIM), New Delhi, India. She has over 6 years of teaching experience in reputed business schools in Delhi NCR. She has taught a number of courses like Management Science, Operations Management, Total Quality Management, Business Statistics, and Project Management to B.B.A., B.M.S., and M.B.A. students. She has received her Bachelor's (Mathematics), Master's (Operational Research), M.Phil. (Operational Research), and Ph.D. (Operational Research) degrees from the University of Delhi, India. She is actively involved in the research on Marketing, Supply Chain, Optimization, and other related areas. She has published several articles in international journals of repute and in national and international conference proceedings.

Dr. Sunil Sharma is a Professor at FMS (Faculty of Management Studies), University of Delhi, Delhi, India. He received his M.Tech. from IIT Delhi and Ph.D. in Total Quality Management from FMS, University of Delhi. He has been teaching, researching, and consulting in the area of Operations Management, particularly Supply Chain Management and Total Quality Management, for the past over three decades at FMS. He received UGC Research Award in 2002 for his work on Supply Chain Management in Indian Industry. He has been involved as a resource person in corporate training and is currently convener of the Management Development Programs (MDPs) at FMS. He is also an active member of the POMS, USA and Vice-President of its India chapter. He attended the International Teachers Program (ITP) at Kellogg School of Management, Chicago in 2010. He also participated in the Global Colloquium on Participant Centered Learning at Harvard Business School, Boston, USA in 2013. He has guided research work for 10 Ph.D. candidates and around 100 research dissertations. He received a Dewang Mehta Award for the best teacher in operations management. He has published three books and a number of research papers. He has presented research papers and chaired technical sessions in at least 25 national and international conferences held at IIM A, IIM B, NITIE Mumbai, University of San Diego, USA, and various SAARC countries. He is also on Research Jury of National Institute of Fashion Technology (NIFT), New Delhi.

Contributors

Vernika Agarwal
Amity International Business School
Amity University
Noida, Uttar Pradesh, India

Sugandha Aggarwal
LBSIM
Delhi, India

Garba Aliyu
Department of Computer Science
Ahmadu Bello University
Zaria, Nigeria

Shruti Ashok
Bennett University
Greater Noida, Uttar Pradesh, India

Bukhari Badamasi
Department of Library Science & Information Science
Ahmadu Bello University
Zaria, Nigeria

Jatinder Bedi
Department of Operations and Decision Science
IIM Kashipur
Kashipur, Uttarakhand, India

Vishal Bhatnagar
Department of Computer Science & Engineering
Ambedkar Institute of Advanced Communication Technologies and Research
Delhi, India

Subhajit Bhattacharya
Accenture

Biswarup Choudhury
LBSIM
Delhi, India

Deepika Dhingra
School of management Bennett University
Bennett University
Greater Noida, Uttar Pradesh, India

Kartik Kakani
LBSIM
Delhi, India

Vaishali Kaushal
Delhi School of Management
Delhi Technological University
New Delhi, India

Maneesh Kumar
Cardiff Business School, Logistics & Operations Management Section
Cardiff University
Cardiff, UK

Snigdha Malhotra
Amity International Business School
Amity University
Noida, Uttar Pradesh, India

Megha Mishra
Shyama Prasad Mukherjee College
University of Delhi
Delhi, India

Upnishad Mishra
Institute of Management Studies
Banaras Hindu University (BHU)
Varanasi, Uttar Pradesh, India

Ashutosh Mohan
Institute of Management Studies
Banaras Hindu University (BHU)
Varanasi, Uttar Pradesh, India

Ishi Mohan
Faculty of Commerce
Banaras Hindu University (BHU)
Varanasi, Uttar Pradesh, India

Sriram Narayanan
Broad College of Business Supply
 Chain Management Department
Michigan State University
Lansing, Michigan

Onawola, H.J.
Department of Software Engineering
 and Department of Information
 Systems
School of Information Technology &
 Computing
American University of Nigeria
Yola, Nigeria

Sidhartha S. Padhi
QMOM Group
IIM Kozhikode
Kozhikode, Kerala, India

R. K. Padhy
Department of Operations and
 Decision Science
IIM Kashipur
Kashipur, Uttarakhand, India

Arvind Panwar
University School of Information
 Communication and Technology
Guru Gobind Singh Indraprastha
 University
Dwarka, Delhi, India

Sandip Rakshit
Department of Software Engineering
 and Department of Information
 Systems
School of Information Technology &
 Computing
American University of Nigeria
Yola, Nigeria

Raju Ranjan
School of Computing Science
 Engineering
Galgotias University
Greater Noida, Uttar Pradesh, India

Aaron Ratcliffe
Walker College of Business,
 Department of Marketing &
 Supply Chain Management
Appalachian State University
Boone, North Carolina

Abhishek Sharma
School of Management
Bennett University
Greater Noida, Uttar Pradesh, India

Manisha Sharma
School of Business Management
NMIMS University
Mumbai, Maharashtra, India

Sunil Sharma
Faculty of Management Studies
University of Delhi
Delhi, India

Tilottama Singh
Amity International Business School
Amity University
Noida, Uttar Pradesh, India

Vidushi Singla
Shyama Prasad Mukherjee College
University of Delhi
Delhi, India

Contributors

Ayush Srivastava
School of Management
Bennett University

Reema Thareja
University of Delhi
Delhi, India

Rajan Yadav
Delhi School of Management, Delhi
  Technological University
New Delhi, India

Nikunj S. Yagnik
Automobile Engineering Department
CVM University
Gujarat, India

1 Understanding the Industry 4.0 Revolution Using Twitter Analytics

Jatinder Bedi and R. K. Padhy
IIM Kashipur

Sidhartha S. Padhi
IIM Kozhikode

CONTENTS

1.1 Introduction 2
1.2 Literature Review 3
1.3 Methodology 5
    1.3.1 Data Collection 5
    1.3.2 Data Analysis 6
        1.3.2.1 Descriptive Analysis 6
        1.3.2.2 Content Analysis 6
        1.3.2.3 Network Analysis 6
        1.3.2.4 Topic Modeling 6
1.4 Analysis and Results 7
    1.4.1 Descriptive Analysis of Tweets 7
    1.4.2 CA of Tweets 9
        1.4.2.1 Sentiment Analysis of Tweets 12
        1.4.2.2 Emotional Scores of Tweets 14
    1.4.3 Network Analytics of Tweets 15
    1.4.4 Topic Modeling of Tweets 16
1.5 Discussions 17
    1.5.1 The Fourth Industrial Revolution Is Characterized by What Tweets? Can We See Any Particular Patterns of Communication and Information Diffusion in These Tweets? 17
    1.5.2 Which Topics and Contents Are Shared in Twitter? Can We Find Any Prevalent Topics or Contents? 18
    1.5.3 What Are the Features of Those Users Who Discuss These Topics Related to Industry 4.0 on Twitter? 19
    1.5.4 What Sentiments Do These Tweets Contain? What Types of Tweets Tend to Contain Sentiment? 19

DOI: 10.1201/9781003140474-1

1.6	Conclusions	21
	1.6.1 Professional Use of Twitter	21
	1.6.2 Organizational Use of Twitter	21
	1.6.3 Stakeholder Engagement	21
	1.6.4 Hiring	21
	1.6.5 Sales Channel	22
	1.6.6 Social Listening	22
	1.6.7 Risk Management	22
1.7	Limitation and Future Research	22
References		23

1.1 INTRODUCTION

In recent times, there has been a trend in the research community where social media (SM) has made a lot of impacts (Aral et al., 2013; Harris, 2014; Kalampokis et al., 2013). SM data provide a vast record of humanity's everyday thoughts, feelings, and actions at a resolution previously unimaginable. Because user behavior on SM is a reflection of events in the real world, researchers have realized that they can use SM to forecast and make predictions. To comprehend the context, we are focusing on Twitter as one of the social networking sites. Sharing contents on Twitter is considered as "open" as anyone can access them, whereas in other social networking sites like Facebook, the content is not open because of user-based restrictions. Thus, this study aims to improve our understanding of the fourth industrial revolution and its associated initiatives in the context of SM. Specifically, the chapter proposes a novel, analytical framework for analyzing the fourth industrial revolutions' tweets. As per our knowledge, a practical methodology or framework for analyzing SM data, related to Industry 4.0 practice, is not reported in the literature. Thus, this research is expected to make important contributions to the Industry 4.0 community. Twitter Analytics (TA) comprises three types of analytics, namely, descriptive analytics (DA), content analytics (CA), and sentiment analytics (SA). This methodology is applied to around 31,000 tweets. These tweets were collected using the hashtags #Industry4.0, #factoroffuture, and #AdvanceManufacturing. The findings are based on four general inquiries:

A. Which tweets characterize Industry 4.0? Is there a particular pattern of communication among these tweets?
B. What are topics shared on Twitter? Were any prevalent topics found?
C. What are the features of the users who use Industry 4.0-related tweets?
D. What are the sentiments of these tweets?

These findings help us acquire greater insights into Twitter's use and play a potential role in understanding the fourth industry revolution and further discussing its research implications.

We have organized the chapter as follows: Section 1.2 presents a brief review of all the associated initiatives of the Industry 4.0 revolution; Section 1.3 presents a review of data analysis methods of Twitter and the use of Twitter in other areas as well as a framework for analyzing Twitter data; Section 1.4 discusses about the application of the TA framework to the analysis of 31,000 tweets and metadata, reports a broad range of intelligence from descriptive, content, and SA techniques, and further discusses them in the context of the fourth industry revolution. Section 1.5 discusses research implications. Sections 1.6 presents the conclusions of this study. Finally, Section 1.7 reports the limitations and possible future research directions.

1.2 LITERATURE REVIEW

Industrial revolution has been evolving from the early days of the human existence, and the manufacturing industry has driven this evolution. The first revolution (Industry 1.0) was a shift from agriculture to the industrial society, and the major contributor of this shift was product volume. There was a large gap between supply and demand during this era, where demand exceeded the supply. Post Industry 1.0, the industry experienced an era of scientific management. This was called Industry 2.0, and it was a period where products were focused on both volume and variety. There were many technological innovations that also happened in different industries, and the major ones included power, electronic and mechanical gadgets, and vehicles. Industry 3.0 (from the 1980s to today) is known for technical innovations. It was more about the shift from simple to digital, particularly in the gadget business. Followed by the third revolution, Industry 4.0 is derived from the term Industrie 4.0 (Kagermann et al., 2013). The fourth industrial revolution has been defined by multiple researchers (e.g., Kagermann et al., 2013) in the context of cyber-physical production systems. The revolution is about taking manufacturing to one level up, in terms of various areas which are part of legacy manufacturing processes. The manufacturing in the era of the fourth industrial revolution also helps us attain a higher level of precision in terms of personalization of the products for the end users. The major change brought in by Industry 4.0 is enabling the production system to be adaptive, analytical, efficient and in synchronization with the systems. Another very exciting feature of the Industry 4.0 production systems is the self-learning capabilities, which is all about learning from the decision errors. The industry 4.0 ecosystem continuously analyzes machine data and ensures that the systems self-learn from the past.

These new manufacturing systems are introducing a new industrial revolution and are referred to as Factory of Future, Industry 4.0, Intelligent Manufacturing, Advance Manufacturing, and e-factory based on the different geographies started by the German government in 2012 (Kagermann et al., 2013) to keep up its global competitiveness in manufacturing industries. Other countries also geared up to adopt Industry 4.0, e.g., South Korea paid attention to Industry 4.0, which Korean government mentioned with an initiative called 'Innovation in Manufacturing 3.0'. It emphasized four propulsion strategies and assignments for a new leap of Korean manufacturing (Kang et al., 2016). The Chinese government strategies like 'Made in China' by 2025 along with the 'Internet Plus' plan are focused toward the manufacturing sector with a clear-cut goal of getting the information potential to the industries in China (Li, 2015). Literature also suggests the upward trend of Industry 4.0 applications in the Japanese manufacturing sector. It goes back to the year 2015, when Japanese government adopted some plans based on Industry 4.0 technologies, with an objective to enable the manufacturing sector as 'Super Smart Society' (Fujii et al., 2018). The Japanese has expressed the need for the e-manufacturing or e-factory even in the past decade, where the use cases of e-maintenance served as the major need for this digital transformation of the conventional factories (Koc et al., 2005). Comparable activities were proposed in the United States by upholding a savvy producing plan (Smart Manufacturing Leadership Coalition, 2011) and suggested connecting everything using IoT (Porter and Heppelmann, 2014). In 2013, the UK government's strategic goals were published in foresight, which presented a long-term vision for its creation sector until the year of 2050, also declared as 'Future of Manufacturing'. The aim was to refocus and rebalance the policy to support the future's manufacturing sector. The European Commission introduced the Public-Private Partnership (PPP) on 'Factories of the Future (FoF)' in 2014. It was under the Horizon 2020 program that intends to give about 80 billion euros of accessible financing over 7 years (from 2014 to 2020) (Liao, et al., 2017). The EU's objective was to achieve a well-organized integration of new varieties of production equipment which are interconnected with each other for the implementation of Factory of Future (Attaran, 1989). This is visualized as the "Factory of Future", and the driving principles are advanced manufacturing, smart manufacturing, and customized smart product within the production environment to support product design, scheduling, dispatching, and process execution throughout factories and production networks to increase efficiency and enable individualization of products (Wang et al., 2016).

Another principle used in the fourth industrial revolution is intelligent manufacturing. The central idea of intelligent manufacturing is to leverage information from a sensor to realize automatic real-time processing similar to intelligent optimized decision-making. Intelligent manufacturing is considered for horizontal integration across associate enterprises' production networks. Similarly, intelligent manufacturing is considered for vertical integration of the enterprise's control and management layers and product life cycle integration, etc. (Shen et al., 2006). Intelligent manufacturing aims to boost product innovation ability, gain fast market response, and enhance automatic, intelligent, versatile, and extremely efficient production processes (Brewer, Sloan, and Landers, 1999). From a strategic point of view, China has shared a Made in China 2015 strategy, which aims at innovation, quality, and efficiency within its manufacturing domain. As we documented, we found this term in China's intelligent manufacturing initiative, which can drive all manufacturing business executions by merging ICT, automation technology, and manufacturing technology (Kennedy, 2015). E-factory is another term relating to the fourth industrial revolution manufacturing process, and this involves using digitization for the same. As the word itself suggests, the e-manufacturing process talks about the optimization of productivity and energy conservation with the help of the internet (Kšksal and Tekin, 2012). As we digitize these factories, they become more visible, things are more measurable, and they can be managed easily. As the factories collect more and more data (generated by machines and collected by sensors), the factories will become more intelligent in decision-making, leading to newer opportunities in manufacturing areas. The e-factory approach's significance is indeed broad: enabling technologies like sensing, smart robotics, automation of knowledge work, IOT, cloud services, 3D printing, etc. (Zurawski, 2016). For research and practice, the Industry

Understanding the Industry 4.0 Revolution

4.0 adoption is critical as the adoption brings value to the manufacturing outputs and systems by integrating cutting-edge technologies in the manufacturing and services. However, there are still a lot of uncertainties either in investments or in unclear benefits, which Industry 4.0 brings to application areas (Kamble et al., 2018).

1.3 METHODOLOGY

1.3.1 Data Collection

TA focuses on the frequency of tweets, association between the tweets, and clustering of tweets for knowledge extraction. Similar activities can be performed employing data pipeline for mining the tweets. Thus, in this study, tweet data collection has been performed multiple times using the framework reported in Figure 1.1 for drawing statistical inferences and knowledge extraction.

Due to unstructured data collected via API services, data analysis becomes challenging. Hence, the application of multiple methods and benchmarks is mandatory to extract information from unstructured Twitter data to get more insightful inferences compared to conventional data (Chau and Xu, 2012). The ideal approach is to pick insights from Twitter's Industry 4.0-related keywords for a given period of time. Considering the volume of tweets that go live each day, we needed an information inspecting process, which will depend on hashtags and catchphrases.

To gather tweets that are related to Industry 4.0, we directed a progression of hashtags, which included 'industry', 'nextindustry', and '4.0'. As a result, we found out that #Industry4.0, #advancemanufacturing, and #factoryoffuture are the most common hashtags used by Industry 4.0 experts. Our formal information was the most recent 1-year data (collected from 18 August 2019 to 17 July 2020), and the dataset incorporated 31,000 tweets with the hashtags related to Industry 4.0. Subsequently, three common methods have been used to extract the information from the unstructured data collected, i.e., DA, CA, and social network analysis (NA) with few more applicable benchmarks.

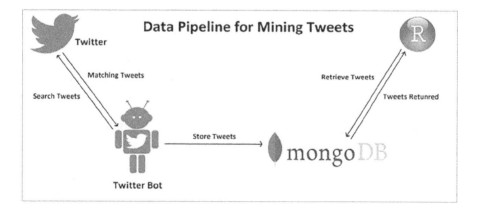

FIGURE 1.1 Data pipeline for mining tweets.

1.3.2 DATA ANALYSIS

1.3.2.1 Descriptive Analysis

DA pays attention to descriptive statistics, such as the frequency of tweets, its distribution, and topics. Summary data techniques are commonly applied in exploratory studies. Tweets are enriched with diversified data, which enable intelligent extraction and facilitate an extensive set of benchmark referencing. Another dimension that is of business value is the source of the tweets, i.e., who is tweeting, replying, or retweeting. These metrics help us identify who is the most active user/group (Bruns and Stieglitz, 2013). This information plays a role in network analytics, which finally helps in popularity analysis. In the end, a significant part of tweets has at least one URL in their raw text. URLs can be classified as release notes from news, outline documents, artifacts, and more.

1.3.2.2 Content Analysis

Twitter information is basically "disordered" in nature, so it becomes compulsory to apply content investigation techniques, viz. the extraction of data from Web 2.0 using strategies like data recovery and content examination (Chau and Xu, 2012). A tweet is an easy-going short content comprising the index of terms, catchphrases, and URL, and it is imperative to carefully refine the content. Along with this, tweets also contain conclusions, for which message examination procedures like assessment investigation have to be carried out to comprehend the feelings of the users about a theme. The content examination pre-refines the unstructured information into a structured form with the assistance of systems like tokenization, n-grams, and disposing of stop words (Weiss et al., 2010). Then this structured information is used for accumulation, most successive words, building content dendrograms, and using different AI-based calculations. This examination encourages us to distinguish the subjects of dialogue, along with the use of online networking like Twitter to Industry 4.0 setting. The subjects work as a contribution to the archive examination strategy like grouping. Hashtags are a basic part of a tweet, and understanding its recurrence and its affiliation enables us to better use the hashtag. This can also help in comprehending ideas in the Industry 4.0 setting over Twitter.

1.3.2.3 Network Analysis

Engagement on Twitter is derived by retweets and replies. Network analysis enables us to remove orchestrated information from tweets using frameworks, which are insightful trains (Burt et al., 2013). This framework showcases the models in the relationship among customers. Distinctive framework estimates give us a quick depiction of such a framework. Hypothesis on the system additionally brings the idea of centrality investigation, which can comprehend as prevalence examination. This investigation will use hub-level measurements as a degree of betweenness centrality that will uncover persuasive hubs in the system.

1.3.2.4 Topic Modeling

While mining the unstructured data, there is a need for a division of information into the common gatherings, which can translate the information independently. Point demonstrating is a strategy for unsupervised order of such unstructured data. It's fundamentally the same as bunching on numeric information, which discovers common

Understanding the Industry 4.0 Revolution

gatherings of things when we don't know what we are searching for. Latent Dirichlet allocation (LDA) is an especially well-known strategy for fitting a theme model (Blei et al., 2003). It regards each report as a blend of points, and every subject as a blend of words. This enables archives to 'cover' each other as far as substance, instead of being isolated into discrete gatherings, such that mirrors commonplace utilization of characteristic language. For example, each archive may contain words from a few points specifically extents. As in a two-subject model, we can say 'Document 1 is 60% of topic A and 40% of topic B, while Document 2 is 25% of subject A and 75% theme of topic B.' Similarly, we could envision a two-subject model of neighborhood news, with one point for 'sports' and the other for 'health'. The most widely recognized words in the theme of the game may be 'player', 'score', and 'team', while the health subject might be comprised of words, for example, 'medicine', 'facilities', and 'patients'. Critically, words can be shared between topics; a word like 'budget' may show up in both similarly. The LDA is a mathematical method for assessing both of these, i.e., finding the blend of words that are related to every point, while additionally deciding the blend of subjects that portrays each document.

1.4 ANALYSIS AND RESULTS

1.4.1 Descriptive Analysis of Tweets

DA of tweets for #Industry 4.0, #AdvanceManufacturing, and #FactoryOfFuture was performed, and the following results were obtained. Let's understand the DA results in the following metrics:

- **Tweet Statistics**: Statistics of around 31,000 tweets, original tweets, retweets, and @reply that represent 43% (13,320), 15.6% (4,862), and 14.06% (4,360), respectively, were studied. Industry 4.0 themes are extremely different, as proven by the number of hashtags (3,326) in the information. Prominent hashtags had all the earmarks of being point regions canvassed in scholastic diaries. They include #IOT, #IIOT, #Big data, #Analytics, #Cybersecurity, #Innovation, #Automation, #digital, #sensors, #cyberattacks, #smart city, #Block chain, #5G, #drone, #robotics, #RPA, #ML, #deep learning, #cloud, #AI, #PlanetEarth, #SubstainableDevelopment, and #Futureofwork (Figure 1.2).

    More than 19,000 tweets (60% of the tweets) contained more than two hashtags, demonstrating that most of the tweets have a cover with different zones of intrigue. For example, tweets contain at least two hashtags: #Industry 4.0, #AI, #MachineLearning, and #Bigdata.
- **User Analysis**: We found 2,847 novel clients in the dataset. This implies that every client puts out: 10.8 tweets, 4.67 unique tweets, 1.68 retweets, and 1.51 @replies per client (Chae, 2015).
- **URL Analysis**: URLs are available in tweets. 20% (6,110) out of all the tweets have at least one URL. 2,533 distinct URLs were obtained from these tweets. Particularly, the general population who are effectively associated with the exchanges use URL in their tweets. As we observed that only 20% of the tweets have the URL presence, it implies that there is still less web content for the topic as compared to other traditional topics (Figures 1.3 and 1.4).

8 Industry 4.0 Technologies for Business Excellence

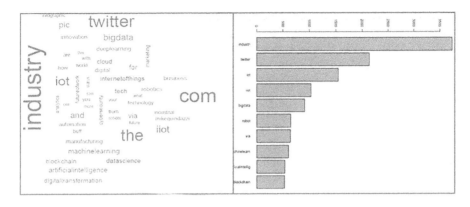

FIGURE 1.2 DA of #Industry 4.0.

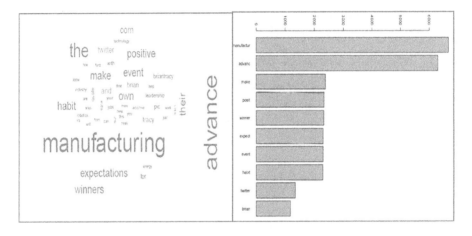

FIGURE 1.3 DA of #AdvanceManufacturing.

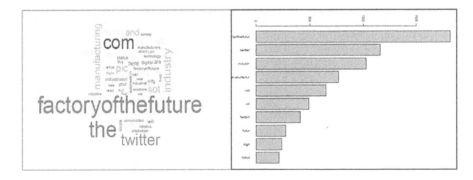

FIGURE 1.4 DA of #FactoryOfFuture.

1.4.2 CA OF TWEETS

The CA mainly talks about word investigation, hashtag examination, and supposition investigation. The word analysis can be categorized into term frequency analysis and clustering (Chae, 2015). Performing the term frequency analysis, the most popular words in the tweets were Industry (4,660), AI (2,980), IOT (2,399), digital (1,965), IoT (1,643), robot (1,034), bigdata (951), machine learning (652), blockchain (530), etc.

Figure 1.5 reports the dendrogram where cluster analysis of #Industry4.0 tweet information has been given. If we move along the vertical axis of the dendrogram and cross the horizontal axis, each line addresses a social affair that was observed while combining the articles into clusters. For example, if we look at 40 on the vertical axis and move along the horizontal axis, we will cross four lines. This describes a four-bunch arrangement, and if we go further down the branches, we can see the

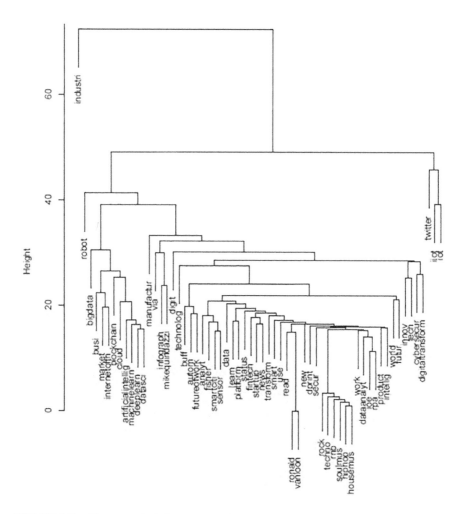

FIGURE 1.5 Cluster analysis for #Industry4.0.

focus words that are grouped under four clusters. Since the vertical axis represents how similar the perceptions were at the time when they were merged, clusters whose branches are close to each other aren't altogether trustworthy.

In the dendrogram, we observe four specific clusters: the right-hand bundle seems to involve two specific clusters of Twitter and IoT, while the second group of the left hand will be gathering a comparable extend for Twitter data. These clusters are mass information, cloud, square chain, and man-made brainpower, which shows how these terms connect each other and are driven by one regular fuel on the information.

As shown in Figure 1.5, we studied #Industry4.0, and Figure 1.6 deciphers clustering yield for #AdvanceManufacturing on Twitter. In this case, if we move along the horizontal axis from 40 on the vertical axis, we will cross four lines. This characterizes a four-group arrangement. By going down the branches further, we see that the words are classified into four clusters. Since the vertical axis represents how close the

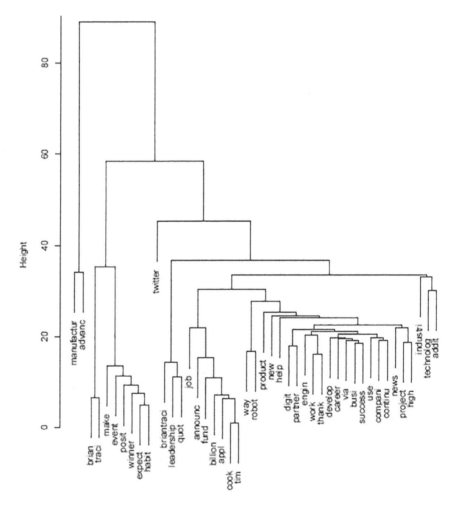

FIGURE 1.6 Cluster analysis for #Advance Manufacturing.

Understanding the Industry 4.0 Revolution

perceptions were to one another, clusters whose branches are very near to each other aren't likely to be entirely solid.

In cases, where there is a major contrast along the vertical axis, between the last merged cluster and the consolidated one, it demonstrates that the groups are most likely working admirably presenting the structure of information. In the dendrogram, there are four particular groups; the right-hand group comprises two clusters of different types of manufacturing techniques, while a large portion of the left-hand cluster groups together a similar range of height for Twitter information. These groups consist of influencers like the brand Traci and indicated how influencers spread awareness about advanced manufacturing.

Figure 1.7 reports the clustering yield for #Factoryoffuture tweets. In this case, if we move along the horizontal axis starting from the point 40 on the vertical axis, we

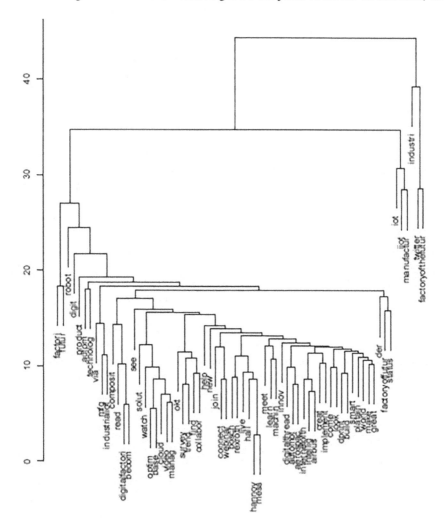

FIGURE 1.7 Cluster analysis for #factory of future.

will cross two lines. This showcases a two-cluster solution. If we move further down the branches, we see words and subjects incorporated into two groups as the vertical axis presents the similarity between the perceptions and the point where the words are merged into clusters.

As the vertical axis represents how close one perception is to the other and the point at which they are merged into clusters, the clusters whose branches are too close to each other are most likely not truly solid. In cases where on the vertical axis there is extreme contrast between the last consolidated group and the one which was combined latest, it demonstrates that the groups formed are working admirably and present the structure of information.

In the dendrogram, there are two particular gatherings: the right-hand bunch appears to comprise industries, manufacturing plants of the future, and Twitter gatherings, while the vast majority of the perceptions in the left-hand bunch are gathering at about a similar tallness go for this tweets information. These gatherings consist of words related to autonomy, digitization, development, and so forth, unmistakably displaying the empowering agents for the plant of the future.

1.4.2.1 Sentiment Analysis of Tweets

Sentiment analysis is also called opinion mining. When we analyze content, we look for passionate words to conclude whether the content is positive or negative. One approach is to break down the content of a tweet and sentiment of the tweet at individual word level (Wickham, 2014). There are several strategies for sentiment analysis, and we have chosen three widely used vocabularies: AFINN (Nielsen, 2011), Bing (Liu, 2012), and NRC (Mohammad and Turney, 2013). Each of these dictionaries depends on unigrams, i.e., single words. These dictionaries contain numerous English words, which are doled out for positive/negative opinions and potential feelings like joy, anger, sadness, etc.

The NRC dictionary classifies words in a parallel design ("yes"/"no") into classifications of positive, negative, anger, expectation, disgust, fear, joy, sadness, surprise, and trust. The Bing vocabulary sorts words in a twofold manner into positive and negative classifications. The AFINN dictionary doles out words with a score that keeps running between −5 and +5, with negative scores showing negative sentiment and positive scores demonstrating positive notions.

As shown in Table 1.1, the positive polarity score for #Industry 4.0 (60%), #AdvanceManufacturing (90%), and #factoryoffuture (60%) outshines the social opinion about the topic. Another dimension to evaluate that we have obtained live data about the #Industry4.0, #Advance Manufacturing, and #factoryoffuture over

TABLE 1.1
Polarity Table

#Industry4.0		#AdvanceManufacturing		#Factoryoffuture	
Polarity Table	Percentage	Polarity Table	Percentage	Polarity Table	Percentage
Negative	14.84	Negative	2.83	Negative	10.03
Neutral	24.39	Neutral	6.37	Neutral	29.90
Positive	60.76	Positive	90.81	Positive	60.07

Understanding the Industry 4.0 Revolution   13

the Twitter sentiment visualization application (Healey and Ramaswamy, 2013). Figures 1.8–1.10 show the perceptual maps for #industry4.0, #advancemanufacturing, and #factoryoffuture. According to the maps, there is a clear inclination toward high confidence for these topics.

Accumulations of tweets are mapped using multidimensional graphical notations. Individual tweets are represented as circles, and each circle's transparency, size, brightness, and color are distinctive based on the general valence of the tweet. Positive tweets are green, while the negative ones are blue. Dynamic tweets are brighter, while repressed tweets are darker. Longer tweets speak to progressively sure gauges, and progressively misty (e.g., less straightforward) tweets represent more confident estimates.

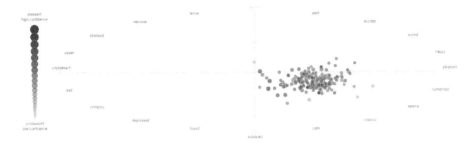

FIGURE 1.8   Perceptual map of #Industry4.0.

FIGURE 1.9   Perceptual map of #AdvanceManufacturing.

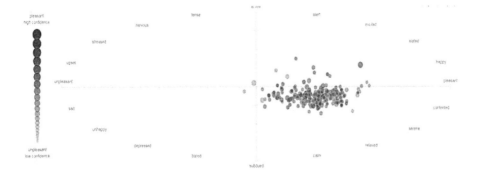

FIGURE 1.10   Perceptual map of #FactoryOfFuture.

1.4.2.2 Emotional Scores of Tweets

In the literature review section, we discussed the barriers encountered by the stakeholders while adopting Industry 4.0. The barriers indicated multiple emotions like fear, joy, surprise, and excitement. To extract such insights from tweets, the emotion analysis was done, and the results are shown in Figures 1.11–1.13.

Positive emotion is an outlier for all the hashtags, and the interesting pattern comes out from the emotion of fear and joy. It shows that there is a fear of the change among the social community; however, the community is also looking forward to this new revolution. It can be interpreted in a way that the part of the community who are happy with this new revolution is more in number as compared to the people who fear it.

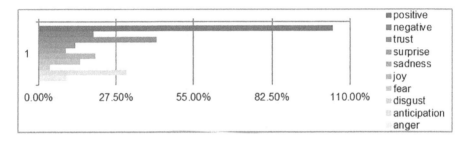

FIGURE 1.11 Emotion score for #Industry4.0.

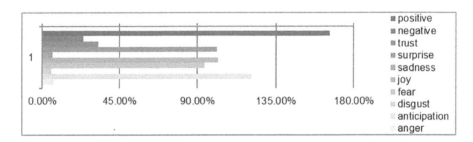

FIGURE 1.12 Emotions score for #AdvanceManufacturing.

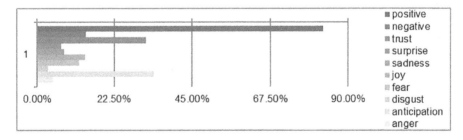

FIGURE 1.13 Emotions score for #FactoryOfFuture.

Understanding the Industry 4.0 Revolution 15

1.4.3 Network Analytics of Tweets

The network graph represents frequent tweets, hashtags, and URLs together with connections between these components. Blue and green nodes represent tweets, orange nodes refer to individuals, yellow nodes present hashtags, and red ones highlight URLs. Bigger size nodes show progressively visiting components. The network graphs for #industry 4.0, #advancemanufacturing, and #factoryoffuture are shown in Figures 1.14, 1.15, and 1.16, respectively.

Critical action takers in a tweet set are identified by the frequency of their tweets, hashtags, and URLs.

FIGURE 1.14 Network graph for #Industry4.0.

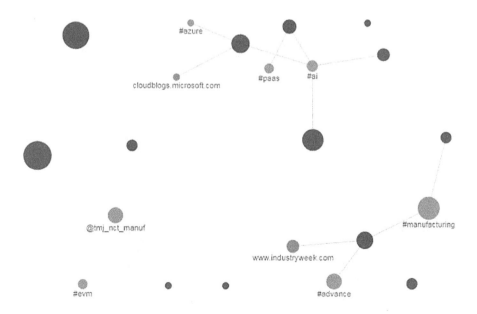

FIGURE 1.15 Network graph for #advance-manufacturing.

16 Industry 4.0 Technologies for Business Excellence

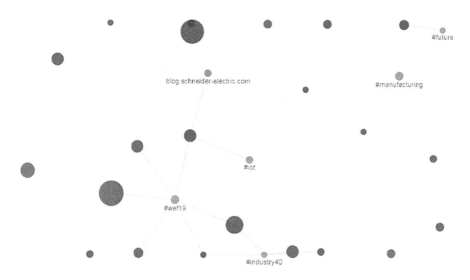

FIGURE 1.16 Network graph for #factoryoffuture.

1.4.4 Topic Modeling of Tweets

Topic modeling helps us to get the common terms between the topics. With the help of the LDA algorithm, we are able to generate a two-topic model as shown in Figure 1.17. Most of the terms like manufacturing, advance, industry, etc. are common in both topics; however, there are few terms like an event, IoT, etc., which are distinct. This gives

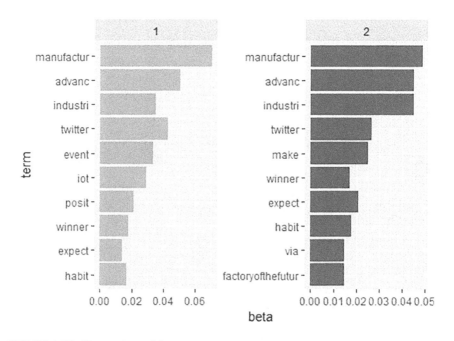

FIGURE 1.17 Two-topic model.

Understanding the Industry 4.0 Revolution

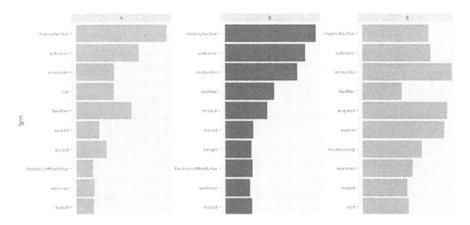

FIGURE 1.18 Three-topic model.

us the insight that Topic 1 is more about the automated data collection; however, Topic 2 is more about the future production systems.

With the intent of studying more about the topics, we created a three-topic model shown in Figure 1.18, and the results were observed to be interesting. We were able to separate the text into more relevant natural groups. Here, Topic 1 talks about the automated data collection, Topic 2 talks about the Industry 4.0 influencers like the brain, and Topic 3 is more about the cyber-physical system technologies like IoT.

1.5 DISCUSSIONS

This section discusses the four questions described in the introduction and our findings regarding the same.

1.5.1 THE FOURTH INDUSTRIAL REVOLUTION IS CHARACTERIZED BY WHAT TWEETS? CAN WE SEE ANY PARTICULAR PATTERNS OF COMMUNICATION AND INFORMATION DIFFUSION IN THESE TWEETS?

- There exist numerous explanations as to why individual and enterprise users use Twitter. Clients disperse information through various ways like using URLs, chat about daily schedules, promote conversations with @reply, and report news about the occasions and occurrences (Java et al., 2007). Research studies show that in an arbitrary example of tweets, the rate of a retweet is simply 3%, and 22% of tweets contain a URL (Boyd et al., 2010). To validate the same, we analyzed our sample. As a result, it was found that 31.4% include a URL and 48.6% include @user.

    This states that the objective is to disseminate information using URL and engage in conversation using @reply. Overall, the data indicate that Industry 4.0 tweets are more conversational and engaging. It is clear by the results that Industry 4.0 experts use Twitter for announcing news and sharing information. Another statistic from a previous study says that 12% of arbitrarily chosen tweets contain a discussion (@reply) (Java et al., 2007). These insight data are somewhat lower than that of #Industry 4.0 tweets (14.06%), making #Industry4.0 tweets seem to be more conversational than open tweets.

The speed of hashtags in #Industry 4.0 hauls consideration as clients incorporate hashtags to show that their tweets are topical, so different people and undertakings with comparable interests can pursue. While it's hard to find a precise solution to the inquiry as to what number of arbitrarily examined tweets would contain no less than one hashtag, the normal answer would be a low rate. Typically, just 13%–23% of tweets by the information innovation engineers contain somewhere around one hashtag (Bougie et al., 2011). Notwithstanding, a great part (35%) of #Industry4.0 tweets contain more than one hashtag. Industry 4.0 subjects are differing, as confirmed by the number of hashtags (3,326) in the information. Most well-known hashtags give off an impression of being like point territories canvassed in scholastic diaries. They include #IOT, #IIOT, #Bigdata, #Analytics, #Cybersecurity, #Innovation, #Automation, #digital, #sensors, #cyberattacks, #smart city, #Block chain, #5G, #drone, #robotics, #RPA, #ML, #deep learning, #cloud, and #AI.

Primarily two variables are playing significant roles in deciding data dispersion of Industry 4.0 tweets. Firstly, the tweets about the convenient issues and difficulties with Industry 4.0 will in general be more noticeably diffused than others. It is especially apparent that an extensive number of tweets containing hashtags such as #Review, #Machinelearning, #worklife, #risk, #sustainabledevelopment, #BigData, #SocialMedia, #SmartSpec, and #SmartContracts are broadly diffused through retweeting. Conversely, work-related tweets are once in a while retweeted. This demonstrates that tweets about the new patterns (e.g., #AI) and issues (#risk, #sustainability) in Industry 4.0 are spread broadly. Secondly, the quantity of hashtags' relationship with the level of dissemination is positive and in all likelihood can be estimated with the retweets. To additionally look at this, we assessed the top retweeted (retweeted more than multiple times) tweets. The outcome demonstrates that those tweets broadly diffused through retweets will in general contain seven hashtags, all things considered. Our discoveries are mostly lined up with investigations of Twitter information in different fields (Stieglitz and Dang-Xuan, 2013; Suh et al., 2010; Yuan et al., 2012). These investigations additionally proposed other potential variables for data dissemination, for example, the number of supporters and following (Suh et al., 2010), and the level of tweet estimation (Stieglitz and Dang-Xuan, 2013).

1.5.2 Which Topics and Contents Are Shared in Twitter? Can We Find Any Prevalent Topics or Contents?

- As explained above, Industry 4.0 subjects are different in the quantity of hashtags. Most mainstream hashtags are like themed territories canvassed in scholarly diaries. They incorporate #IOT, #IIOT, #Bigdata, #Analytics, #Cybersecurity, #Innovation, #Automation, #digital, #sensors, #cyberattacks, #smart city, #Block chain, #5G, #drone, #robotics, #RPA, #ML, #deep learning, #cloud, and #AI. Few noticeable ones were #PlanetEarthFirst,

#SustainableDevelopment, and #Futureofwork. They were also ranked high in the retweets. There were discussions in tweets, which lead toward the digitized synergies which in turn lead toward the optimum process operations. Hence, the optimized use of resources becomes critical here. There are discussions about the challenges to manage the growth, while there will be transition toward Industry 4.0. Challenges will also occur in the energy supply department.

1.5.3 WHAT ARE THE FEATURES OF THOSE USERS WHO DISCUSS THESE TOPICS RELATED TO INDUSTRY 4.0 ON TWITTER?

- Our research uncovers significant data that reveal the dynamic in Twitter. Some of the clients are unemployed, while others are professionals. About all the tweets under this group of clients are unique, and they are expert in innovation-related jobs. Another group of clients incorporates industry professionals. Other dynamic group includes magazines (AMS magazine), IT administration organizations (e.g., Tampa), and counseling companies (e.g., Mckinsey). Effective clients get some @reply (or specifies), and their tweets are retweeted. Exceedingly effective clients originate from different areas (e.g., automation and manufacturing) and roles (e.g., automation specialists, online magazines, proficient associations, and IT organizations). Some solution providers, manufacturers, and retailers are visible discussing about the artificial intelligence, big data, internet of things, machine learning, etc. However, majority of the clients are not exceptionally dynamic in #Industry 4.0 tweets. Past researchers demonstrate that a low number of clients represent a huge segment of tweets.

  For instance, a past investigation where over 11 million tweets were abridged discovered that just 5% of clients represent 75% of tweets (Cheng et al., 2009). Additionally, when an arbitrary example of around 3,000 tweets was outlined by a couple of specialists in 2009, it was discovered that around 10% of clients represent over 90% of tweets for the vast majority of the themes (Heil and Piskorski, 2009). These examples further validate our research on Industry 4.0 tweets. Just 2% (478 clients) and top 8% (2,769 clients) represent very nearly 45% and 36% of tweets, respectively. This is increasingly clear in the proportion of clients and unique tweets. The top 10% of clients represent 55% and 35% of unique tweets. In any case, the other 90% of clients represent an expansive portion (45%) of retweets. Indicating this gathering of clients will be supporters of those exceedingly effective clients in the network.

1.5.4 WHAT SENTIMENTS DO THESE TWEETS CONTAIN? WHAT TYPES OF TWEETS TEND TO CONTAIN SENTIMENT?

- The tweet data test for #Industry4.0 consists of moderately positive feeling. This finding was likewise backed up by research. Rather than the general tweets, which are either in support of any political plans or media outlet, most Industry 4.0 tweets were about occasions, fabricating capacity news, business use case, changes, reports, innovations, abilities, and administrations notices.

Likewise, it was observed that separated from the positive estimation, even impartial slant was additionally more than the negative assessment among the tweets. This suggests that a portion of the clients and ventures are hanging tight to see the uses of Industry 4.0 on business, devices, and innovation which are not infiltrated into the assembling or different areas. Among the tweets, those identified with hazardous activities, assembling, and security are viewed as more nostalgic than others. Discernibly, hazard-related tweets will in general convey more negative opinions than those of different points. There have been tweets talking about the major job risk, which this disruption brings along with the other benefits. There appears to be excessively increasingly negative tweets in the point of assembling (Table 1.2).

TABLE 1.2
Twitter's Potential Role for Industry 4.0 Professionals and Companies

Examples: Professional Use

In Learning: (1) AI, ML, and automation are few topics on which experts are being usually followed; (2) looking for different topics (or keywords)

In Promoting: Tweeting/retweeting useful data, information, opinions, and conclusions on timely subjects and issues

In Networking: Using some Twitter features such as @reply and followers/following

Examples: Organizational Use

In Stakeholder engagement: (1) Twitter is often considered as a communication platform; (2) spreading positive pictures as socially capable and successful organizations; (3) connecting with a large public with examples of success stories through tweets and retweets

In Hiring: (1) Tweets are made for different job openings; (2) mining Twitter through user timelines can help identifying talented professionals using different kinds of analytics like descriptive, network, and CA

In Demand Shaping and Sales: (1) Twitter as a channel for sales; (2) tweet about production information and lead to encourage followers to retweet the same; (3) creating and identifying customer demands by tweeting about offers.

In Social Listening for New Product/Service Development: (1) Twitter used as a social sensor; (2) using sentiment analysis to extract demand signals from the customers and the markets; (3) getting feedbacks, ideas, and product experience from customers based on the product's quality and service

In Risk Management: (1) Twitter as an event observing and monitoring and collaboration device; (2) keeping a check on relevant events, disruptions, and various other news in real time related to Industry 4.0; (3) broadcasting Industry 4.0 occasions and dangers progressively to different partners in manufacturing area and helping coordinated effort

1.6 CONCLUSIONS

In line with the potential role of Twitter for Industry 4.0 practice and research, underneath we present further ramifications of Twitter for users like experts, associations, and Industry 4.0 researchers.

1.6.1 Professional Use of Twitter

The professional world has been using Twitter for various reasons (Conway et al., 2013). The results show that Industry 4.0 tweets are more engaging and conversational than normal tweets. For experts, Twitter is a basic wellspring of finding current news, occasions, and new information. Contrastingly, Twitter over other media, like pages, books, and meetings, has the key advantage of microblogging in which the client doesn't need to be proactive to extract data when contrasted with other media.

Twitter allows a substantial amount of data pursuit, securing quicker and progressively advantageous. For example, straightforward hunt using 'examination' at Twitter gives sufficient chance to learn the most recent point through pertinent news, recordings, and cheat sheets and finding different experts/scientists and ventures in this rising territory. In addition to learning, Twitter is also used for promotion and networking. Various academicians and a lot of journals also use Twitter for promoting their latest issues and content (Holmberg and Thelwall, 2014). Additionally, Twitter can be an exceptionally compelling device for task experts to advance themselves, for example, through tweets/retweets of applicable and helpful data and supposition on auspicious subjects and issues.

1.6.2 Organizational Use of Twitter

The research findings from #Industry4.0 show that Twitter is widely used for stakeholder engagement, company promotion, and hiring.

1.6.3 Stakeholder Engagement

CA clearly shows that for some infrastructure solution providers, drone service providers, logistics providers, and manufacturers, the use of Twitter is strong in the area of stakeholder engagement. The attention has been on spreading positive image as socially capable and effective organizations. There has been a pattern of connecting with an expansively open audience and overcoming adversity through tweets and retweets.

1.6.4 Hiring

Again, while analyzing the tweet data, there were few tweets where users posted the skills for jobs in the Industry 4.0 domain. Hence, it is very evident that Twitter is used as a hiring tool for most enterprises. The companies are using their Twitter handles to share the job openings and descriptions. Another research method is to track the Twitter client course of events and recognizing skilled experts using clear, substance, and network analytics techniques.

1.6.5 SALES CHANNEL

While examining the data, it was seen that a lot of enterprises (e.g., Logic Bay, salesforce, and so forth) have shared capacities regarding relocating to new advancements of Industry 4.0. The tweets sharing the creation data have been viewed as a typical pattern, and they have urged the adherents to retweet the equivalent. The aim is to make client requests by tweeting about the offers. Subsequently, Twitter has been used as a business channel with a target to produce more leads for the business. For instance, 'Are your items and administrations wise enough for your clients? Find our key proposals for expanded #CustomerValue and steadfastness in the blog. #SmartProducts #Industry40 #IoT (http://bit.ly/2A1e1qC pic.twitter.com/e11FlXgJxd).'

1.6.6 SOCIAL LISTENING

The companies are using the Twitter platform as a social sensor, which means that the companies are consistently analyzing the Twitter feeds to come up with the new product or service, which are customer centric. This helps in extracting demand signals from clients and markets using strategies like sentiment mining. This also helps in getting customer aspirations and feedback about the existing products and services.

1.6.7 RISK MANAGEMENT

Risk management is an important aspect of any business. So, risk monitoring becomes very critical for every business to sustain. Regular monitoring of internal processes and business helps organizations to keep a check on the risk, limiting the mitigation to internal processes. In case of risk to business from external factors like competition, the latest disruptions, etc., Twitter can serve as an event monitoring and collaboration tool (Chae, 2015). Twitter can help in communicating Industry 4.0 occasions and dangers progressively to different accomplices in assembling zone and helping the coordinated effort.

1.7 LIMITATION AND FUTURE RESEARCH

It is crucial for operation professionals and organizations (and even scientists) to successfully use information gathered from online stages for their operational leverage. The task club has not been that dynamic in recognizing the potential, which internet-based life information brings for them. This paper has outlined the flow circumstance by proposing an explanatory structure for Twitter data. This research has confinements, especially with the information-gathering part. As business 4.0 has been drifting late from the most recent years, so more information is needed for a couple of more years to build a complete picture. As mentioned earlier, we picked #Industry4.0, #AdvanceManufacturing, and #FactoryOfFuture in pursuit of information.

Another thorough methodology uses various catchphrases (e.g., computerized reasoning, square chain) and/or hashtags (e.g., #manufacturing, #IoT, #IIoT) in information accumulation. This methodology would empower using an expansive amount of Industry 4.0 related to Twitter information.

Likewise, this vast volume of tweets which are valuable for big business with Industry 4.0-related choices may not contain such watchwords, for example, activities, computerization, dangers, and so forth. An answer for remediating these situations is to get Twitter information using a vast rundown of catchphrases. We accept that there is a need to upgrade the comprehension of Twitter information in the Industry 4.0 setting. This gives us two headings to tail, one of them is to create nitty-gritty and pragmatic rules, which can help organize tasks to structure Industry 4.0 applications using Twitter. Research association programs with inventive associations can assume significant jobs in this framework.

In this line, activity scientists should initially think about the inventive use of internet-based life in different controls. The other region is concentrating on the effect of SM investment (e.g., advancements, information researchers, and information repositories) on activity execution. This sort of research is earnestly required, as there is expanding regard and enthusiasm for mass information and online networking from ventures and the scholarly community.

REFERENCES

Aral, S., Dellarocas, C., & Godes, D. (2013). Introduction to the special issue—Social media and business transformation: a framework for research. *Information Systems Research*, *24*(1), 3–13.

Attaran, M. (1989). The automated factory: Justification and implementation. *Business Horizons*, *32*(3), 80–85.

Blei, D. M., Ng, A. Y., & Jordan, M. I. (2003, January). Latent Dirichlet allocation. *Journal of machine Learning Research*, *3*, 993–1022.

Bougie, G., Starke, J., Storey, M., & German, D. (2011). Towards understanding Twitter use in software engineering: Preliminary findings, ongoing challenges and future questions. In: *Proceedings of the 2nd International Workshop on Web 2.0 for Software Engineering*. ACM, Waikiki, HI, pp. 31–36.

Boyd, D., Golder, S., & Lotan, G. (2010). Tweet, tweet, and retweet: conversational aspects of retweeting on Twitter. In: *Proceedings of the 43rd IEEE Hawaii International Conference on System Sciences*. Kauai, HI, pp. 1–10.

Brewer, A., Sloan, N., & Landers, T. L. (1999). Intelligent tracking in manufacturing. *Journal of Intelligent Manufacturing*, *10*(3–4), 245–250.

Bruns, A., & Stieglitz, S. (2013). Towards more systematic Twitter analysis: Metrics for tweeting activities. *International Journal of Social Research Methodology*, *16*(2), 91–108.

Burt, R. S., Kilduff, M., & Tasselli, S. (2013). Social network analysis: Foundations and frontiers on advantage. *Annual Review of Psychology*, *64*, 527–547.

Chae, B. K. (2015). Insights from hashtag# supplychain and Twitter analytics: Considering Twitter and Twitter data for supply chain practice and research. *International Journal of Production Economics*, *165*, 247–259.

Chau, M. C. L., & Xu, J. (2012). Business intelligence in blogs: Understanding consumer interactions and communities. *MIS Quarterly*, *36*, 1189–1216.

Cheng, A., Evans, M., & Singh, H. (2009). Inside twitter. [Online]. http://www.sysomos.com/insidetwitter.

Coalition, S. M. L. (2011, June). Implementing 21st century smart manufacturing. In: *Workshop Summary Report*. University of California Los Angles.

Conway, B. A., Kenski, K., & Wang, D. (2013). Twitter use by presidential primary candidates during the 2012 campaign. *American Behavioral Scientist*, *57*(11), 1596–1610.

Fujii, T., Guo, T., & Kamoshida, A. (2018, August). A consideration of service strategy of Japanese Electric manufacturers to realize super smart society (SOCIETY 5.0). In: *International Conference on Knowledge Management in Organizations*, pp. 634–645. Springer, Cham.

Harris, D. (2014). 3 Lessons in Big Data from the Ford Motor Company. https://gigaom.com/2014/02/08/3-lessons-in-big-data-from-the-ford-motor-company/

Healey, C. G., & Ramaswamy, S. S. (2013). https://www.csc2.ncsu.edu/faculty/healey/tweet_viz/tweet_app/.

Heil, B., & Piskorski, M. (2009). New Twitter research: Men follow men and nobody tweets. *HBR Blog Network*. http://blogs.hbr.org/2009/06/new-twitter-re search-men-follo/ (accessed 04.01.13).

Holmberg, K., & Thelwall, M. (2014). Disciplinary differences in Twitter scholarly communication. *Scientometrics*, 1–16. doi.:10.1007/s11192-014-1229-3.

Java, A., Song, X., Finin, T., Tseng, B., 2007. Why we Twitter: Understanding microblogging usage and communities. In: *Proceedings of the 9th WebKDD and 1st SNA-KDD 2007 Workshop on Web Mining and Social Network Analysis*. ACM, New York, pp. 56–65.

Kagermann, H., Helbig, J., Hellinger, A., & Wahlster, W. (2013). Recommendations for implementing the strategic initiative Industrie 4.0: Securing the future of German manufacturing industry; final report of the Industrie 4.0 Working Group. Forschungsunion.

Kalampokis, E., Tambouris, E., & Tarabanis, K. (2013). Understanding the predictive power social media. *Internet Research*, 23(5), 544–559.

Kamble, S. S., Gunasekaran, A., & Sharma, R. (2018). Analysis of the driving and dependence power of barriers to adopt Industry 4.0 in Indian manufacturing industry. *Computers in Industry*, 101, 107–119.

Kang, H. S., Lee, J. Y., Choi, S., Kim, H., Park, J. H., Son, J. Y.,… & Do Noh, S. (2016). Smart manufacturing: Past research, present findings, and future directions. *International Journal of Precision Engineering and Manufacturing-Green Technology*, 3(1), 111–128.

Kennedy, S. (2015). *Made in China 2025*. Center for Strategic and international Studies, Waterloo, ON.

Koc, M., Ni, J., Lee, J., & Bandyopadhyay, P. (2005). *Introduction to e-Manufacturing*. The Industrial Information Technology Handbook, CRC Press LLC.

Kšksal, A., & Tekin, E. (2012). Manufacturing execution through e-Factory system. *Procedia CIRP*, 3, 591–596.

Li, K. (2015). Made in China 2025. Report, Beijing, China.

Liao, Y., Deschamps, F., Loures, E. D. F. R., & Ramos, L. F. P. (2017). Past, present and future of Industry 4.0—A systematic literature review and research agenda proposal. *International Journal of Production Research*, 55(12), 3609–3629.

Liu, B. (2012). Sentiment analysis and opinion mining. *Synthesis Lectures on Human Language Technologies*, 5(1), 1–167.

Mohammad, S. M., & Turney, P. D. (2013). *NRC Emotion Lexicon*. National Research Council, Ottawa, ON.

Nielsen, F. Å. (2011). A new ANEW: Evaluation of a word list for sentiment analysis in microblogs. arXiv preprint arXiv:1103.2903.

Porter, M. E., & Heppelmann, J. E. (2014). How smart, connected products are transforming competition. *Harvard Business Review*, 92(11), 64–88.

Shen, W., Hao, Q., Yoon, H. J., & Norrie, D. H. (2006). Applications of agent-based systems in intelligent manufacturing: An updated review. *Advanced Engineering Informatics*, 20(4), 415–431.

Stieglitz, S., Dang-Xuan, L., 2013. Emotions and information diffusion in social media – sentiment of microblogs and sharing behavior. *Journal of Management Information Systems*, 29, 217–248.

Suh, B., Hong, L., Pirolli, P., & Chi, E. H. (2010). Want to be retweeted? Large scale analytics on factors impacting retweet in Twitter network. In: *Proceedings of the 2010 IEEE Second International Conference on Social Computing*. IEEE Computer Society, Minneapolis, MN, pp. 177–184.

Wang, S., Wan, J., Li, D., & Zhang, C. (2016). Implementing smart factory of Industrie 4.0: An outlook. *International Journal of Distributed Sensor Networks, 12*(1), 3159805.

Weiss, S. M., Indurkhya, N., Zhang, T., & Damerau, F. (2010). *Text Mining: Predictive Methods for Analyzing Unstructured Information*. Springer Science & Business Media, Heidelberg.

Wickham, H. (2014). Tidy data. *Journal of Statistical Software, 59*(10), 1–23.

www.iiconsortium.org.

www.smartmanufacturingcoalition.org.

www.whitehouse.gov/sites/default/files/microsites/ostp/amp_final_report_annex_1_technology_development_july_update.pdf [viewed 15.09.15].

Yuan, T., Achananuparp, P., Lubis, I.N., Lo, D., Ee-Peng, L., 2012. What does software engineering community microblog about? In: *Proceedings of the 9th IEEE Working Conference on Mining Software Repositories (MSR)*. IEEE, Zurich, Switzerland, pp. 247–250.

Zurawski, R. (2016). *Integration Technologies for Industrial Automated Systems*. CRC Press, Boca Raton, FL.

2 The Role of Universal Product Coding (UPC), Global Data Synchronization Network (GDSN) and Product Category Management in Efficient Consumer Response (ECR)

Sunil Sharma
University of Delhi

CONTENTS

2.1 Background of ECR 28
    2.1.1 ECR as a Profit Center in Supply Chain 28
    2.1.2 ECR as a Cost Alleviator in Supply Chain 29
    2.1.3 ECR and Operational Efficiency in Supply Chain 29
    2.1.4 ECR as Risk Mitigator in Supply Chain 30
    2.1.5 ECR and the Inventory Management in the Supply Chain 30
    2.1.6 ECR as a Supply Chain Strategy 31
2.2 Framework of a Global ECR Scorecard 31
2.3 Prerequisite Initiatives for Participation in ECR 32
2.4 Mechanics of ECR 35
2.5 Special Application of ECR in Product Category Management 36
2.6 Universal Product Code (UPC) 38
    2.6.1 Features of UPC Barcode 39
2.7 Overview of GDSN 39
    2.7.1 Product Information Sharing through GDSN and Impact on ECR 41
2.8 ECR in India 42
    2.8.1 An Example of ECR Initiative in India 42

DOI: 10.1201/9781003140474-2

    2.8.2   Workgroups in ECR India 42
    2.8.3   The Future of ECR in India 43
2.9   ECR in World 43
2.10 Conclusion 44
References 44

2.1 BACKGROUND OF ECR

Efficient consumer response (ECR) emerged in the early 1990s in the USA to break down nonproductive barriers in business. ECR emanated from the threats of substitute stores and their supply chain. Globally distributed supply chains and consumer demands focused on better products and services together with intense competition, proliferation in product variety, and developments in information technology, which necessitated the need to develop effective ways of delivering products and services faster and cheaper. ECR is considered as a concept that ensures the grocery chain works smoothly and proficiently in serving the consumers' wants (Perona, 2004).

There are four initiatives that define ECR. These include efficient store assortment, efficient promotion, efficient product introduction, and product replacement (Kurt Salmon Associates, 1993).

  i. Efficient Store Assortment: It enhances the efficiency in use of storage space by reducing duplicate merchandise or proper assortment planning for the store.
  ii. Efficient Promotion: This is achieved by executing a better alternate promotion plan or mix.
  iii. Product Introduction: It entails the new product development (NPD) and promotion of new products/variants in the market.
  iv. Product Replacement: It is done through a replacement of the nonmoving or defective products.

2.1.1 ECR as a Profit Center in Supply Chain

The efficient flow of information between partners as a way of collaborative planning, forecasting, and replenishment ensures that there develops a strategic alliance between the supply chain members. The good relationship and trust thus developed between the supply chain partners at different levels motivates them to improve financial performance so as to share rewards on a win-win rather than win-lose basis. This is achieved due to shorter cycles or frequent exchange of information. These strategies enhance operational efficiency in the chain management which increases profit (Brown, 2001). The ECR goals aim to diminish the supply chain costs that do not hold any consumer value. ECR leverages the competencies of the supply chain to reduce costs while streamlining the operations. ECR has resulted in an increase in profit as it provides a platform for efficient working relationships among the partners in the supply chain.

Another impact of ECR is its ability to retain only the value adding processes of the product after re-engineering, making them the best. ECR is thus composed of a mix of strategic initiatives, operational programs, and a number of enabling technologies that improve its effectiveness. Also, products that are not performing well in the market are phased out and replaced with other products so as to improve the quality and value of the goods supplied, thereby augmenting the market share leading to increase in profit.

2.1.2 ECR as a Cost Alleviator in Supply Chain

The ECR program is supported by a number of technologies in cost reduction in supply chain management, the key of these being barcode/EPC (electronic product code) scanning, electronic data interchange (EDI), computer-aided ordering (CAO), cross-docking, and activity-based costing (ABC). The goal of CAO is to speed up store replenishment process and ensure that products are ordered on time.

Use of bar code/EPC and scanning has become inevitable in the retail industry. Development of global trade item numbers (GTINs) has helped it globally. The main benefits of CAO are labor savings, dependability, and inventory reduction (Griffin, 1998). The reduction in labor requirement enables the supply chain to reduce the production cost as it employs fewer employees (King, 1996). EDI link, in addition to usual exchange of information about ordering, sales, and inventory, facilitates exchange of local and international commercial documents, e.g., related to INCO (International Commercial terms) or delivery terms in a machined processed form, hence reducing the transportation cost.

In cross-docking, products are delivered to warehouses on a continual basis where they are sorted, repackaged, and distributed to different stores without sitting in inventory, crossing one dock to another in 24–48 hours, thereby also facilitating 'product mixing' function.

ABC ensures that all costs are linked to their respective activities. Through ABC, the costs of various activities are estimated and apportioned. This enables the managers to determine costly activities and develop appropriate methods of reducing these costs.

2.1.3 ECR and Operational Efficiency in Supply Chain

ECR enhances the operational efficiency between channel members, reduces time between billing and payment, aids in establishing a paperless system, and eliminates excess costs (Hall & Hall, 1990). The sharing of the information in the supply chain enables the companies to have only efficient inventory levels.

ECR ensures that the manufacturers, wholesalers, and retailers work together as partners, and this reduces the total operating cost, inventories and physical resources that are required during the distribution process. The inventory strategy, which is a part of ECR, enables the efficient delivery of better raw materials that results in production of high-quality products which in turn satisfy the consumer (Coopers, 1998). ECR also ensures that there is smooth flow and traffic of goods in a warehouse with a minimal disruption and re-handling of goods. ECR also plays a significant role in reduction of labor, thereby enabling the supply chain to gain profit.

The CAO, an important component of ECR, enhances efficiency and the smooth movement of goods from suppliers to consumers. CAO is an effective way of meeting consumer needs as orders are well edited and properly stored. The information in the product catalogues used is usually electronic in nature, and this enables faster communication through the modern methods of information transfer such as the internet and mobile phones. The trade partners are able to share first-hand information faster, regardless of the part of the world they may be coming from. This leads to the increase in efficiency of the supply chain and encourages interoperability. Operational efficiency in the supply chain through ECR is also ensured by practicing product category management. Product category management enhances an interactive business process whereby the retailers and manufactures work together to manage the supply chain (Arango, 1999).

2.1.4 ECR AS RISK MITIGATOR IN SUPPLY CHAIN

ECR facilitates production of high-quality products to consumers and hence reduces the risk of reluctant (customers) for buying the products (Kurnia & Johnston, 2001). ECR ensures that the products that reach the consumer are of high quality, fresh, and in time.

The inventory-focused processes in ECR bring about efficient delivery of better raw materials that result in production of high-quality products which in turn satisfy the consumer. This is crucial as it ensures that the customers receive products 'on time and in full (OTIF)', and any complaint from them is settled to their satisfaction.

The positive handling of the consumer feedback about the products and services helps the supply chain performance improve. ECR requires a smooth working relationship among the supply chain members, thereby reducing the risk of miscommunication and mistrust.

2.1.5 ECR AND THE INVENTORY MANAGEMENT IN THE SUPPLY CHAIN

Continuous Replenishment Program (CRP), another important component of ECR, has been used in controlling and monitoring the goods movement from the manufacturer to the distributor (Greenbaum, 1997). CRP reduces the distributor's inventory costs. CRP enhances partnership practices among the members of the distribution channel by allowing products to move from manufacturer to customers (Ellram & Weber, 1989). This leads to a decrease in the inventory cost. CAO's objective is to speed up the store replenishment process and ensure that products are ordered and hence delivered on time, thereby also resulting in better inventory management.

The sharing of (inventory related) information in ECR helps in strengthening the efficiency in supply chain management (Holms, 1998). ECR thus eliminates the extra supply chain costs as well as improves synchronization among demand, production, and distribution over a time frame.

The inventory strategy processes involved in the supply chain enable efficient delivery of better raw materials that result in production of high-quality products, which in turn satisfy the consumer (Brockman& Morgan, 1999). Cross-docking enhances flow of products at the distribution center to the receiving center. Cross-docking ensures the movement of goods in a warehouse with a minimal disruption and handling of goods. The method decreases the inventory cost (Brown, 2001).

The Role of UPC and GDSN in ECR   31

Barcodes and scanners are used to capture the information about an item number, and this aids in tracking the movement of goods and ensuring online visibility and hence help the businesses track and trace the movement of all the items.

2.1.6 ECR AS A SUPPLY CHAIN STRATEGY

ECR links the manufacturers, retailers, and vendors to work together seamlessly in achieving their objective goals and hence reduces mistrust among the partners. ECR results in an unswerving flow of products from/to the distribution center and the receiving/dispatching shipping center. This helps in obviating the need for additional handling and storage in the distribution cycle, thus re-orienting the supply chain.

Most companies strategize to make their supply chain efficient in reducing costs and cycle times by using electronic data network and flows mainly using a Global Data Synchronization Network (GDSN) as an enabler for ECR. These strategic initiatives greatly help in achieving budgetary goals and requirements which are essential for financial efficiency. ECR improves product flow by optimizing inventory levels across the supply chain and making cash flow faster. Total cycle times, warehousing costs, transit losses, and inventory holding periods are reduced, and inventory turnover ratios and order fill rates are increased.

2.2 FRAMEWORK OF A GLOBAL ECR SCORECARD

A global ECR scorecard is a capability assessment tool for fast-moving consumer goods (FMCG) companies for efficient customer response. It comprises four focus areas namely, demand management, supply management, enablers, and integrator as shown in Figure 2.1.

Under demand management, companies can assess demand capabilities on strategic intent and direction of consumer value, people, and organization and information management. It also involves optimized planning and execution of processes for the same. It also entails planning of assortments, new product introductions, and promotions. It also aims to manage consumer knowledge and develop different solutions and channels for consumers.

Under supply management, companies can evaluate supply capabilities considering the strategic direction, people, and organization and information management. A responsive and automated replenishment system together with automated store ordering and transport optimization is also a part of this module. Operational excellence in supplies, production, and distribution together with demand-driven integration is a must in this module.

Enablers include elements like product and shipment identification, electronic data alignment, and communication. Cost, profit, and value can be measured on ABC and consumer value management. Integrators help the companies to bring about CPFR (Collaborative Planning Forecasting and Replenishment) and e-business solutions.

Weight factors or points on each of the above are evaluated for both retailers and suppliers. Scores can then be assigned to assess the strength of the areas in a company. Need gaps can thus be identified and highlighted for a suitable supply chain intervention.

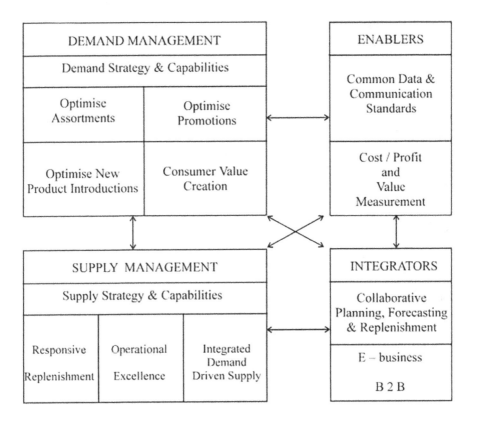

FIGURE 2.1 Framework of Global ECR scorecard. (Adapted from www.globalscorecard. net as retrieved from 10 August 2008.)

2.3 PREREQUISITE INITIATIVES FOR PARTICIPATION IN ECR

The following initiatives need to be executed for fruitful participation in ECR:

i. CRP
ii. CAO
iii. Flow-through distribution
iv. ABC
v. Category management
vi. Integrated EDI

Let us discuss each of these briefly.

i. CRP: It refers to a vendor-managed inventory set-up in which either the vendor continuously monitors a customer's inventory or customer supplies current inventory data so that the vendor makes timely shipments to maintain customer's inventory as agreed to upon levels. It thus results in optimization of the logistics chain by replacing the push or schedule-based inventory replenishments with pull or demand/order-based replenishments. CRP is a method of replenishing products in real time as needed only for the sold amount. It is a concept that supports the ECR strategy for supply chain management of processed foods/groceries. Unlike the MRP (Material Requirement Planning) system, here, the time buckets creating a schedule for certain time periods do not exist. Instead, the products are replenished only for the sold amount as needed in real time, and there is no specific calculation of an order size. However, due to data processing time and economy reasons, complete CRP is difficult to execute, and a method focusing on speed and cost of goods flow is developed. In this method, CRP is used in combination either with the ordering point method in which an economic amount of order is calculated for order placement when inventory reaches a certain RoP (Reorder Point) or with the replenishment ordering system in which it is aimed to return the inventory level to a basic minimum level.

ii. CAO: According to an estimate by FMI (Food Marketing Institute) in the USA, around 47% of all stock-outs are as a result of improper store ordering and forecasting. The Grocery Manufacturers of America (GMA) found the average level of out of stock per cent at typical supermarkets in the USA to be 7.4%. Further, according to US Capgemini, 40% of customers postpone their purchase or buy the product elsewhere after facing a stock-out that results in a 3% loss in sales at a store. The supermarkets are forced to keep 3–5 days of extra inventory even to keep customer service levels of 97%.

All this needs a computer-assisted ordering application, CAO, the key prerequisite for which is a PI (Perpetual Inventory) System. PI is a method for tracking and knowing the value of inventory and quantity of merchandise on hand at any time by tracking sales returns, receipts, and deliveries on the information system. Thus, CAO in conjunction with a PI system would assist store managers by suggesting a product replenishment order based on the store's actual PoS movement history for that item and assured actual on-hand inventory. CAO is driven through direct feeds from the PoS sales data, which, based on specific parameters as listed below, would replenish the store.
 i. Desired shelf inventory
 ii. Reorder quantities/pack-size and their limits
 iii. Minimum and maximum stock levels
 iv. Reorder point (RoP)
 v. Product promotion data
 vi. Delivery frequency
 vii. Replenishment lead time

CAO allows retailers to buy what they sell instead of selling what they buy. The focus is on stocking fast-moving goods and not the slow-moving goods. The average inventory investments following CAO have been reported to be decreased by 15%.

iii. **Flow-through Distribution**: It is also known as pool distribution, and it combines economy of scale with the flexibility of vendor-managed inventory. It is the process of controlling inbound shipments, sorting them by delivery destinations and then sending them out the same day. This process eliminates the need for warehouses and reduces the need for high inventory levels and the time to get the product to the market. In a way, it is a modified form of *cross-docking* only except that the flow-through distribution system also offers other value-added services like *pick and pack* and *kitting and crating*. Flow-through centers would use well-augmented IT services at *hubs* to make shipments move faster.

The pool distribution can be used by retailers and manufacturers both, whereby it would make the movement of products faster through the supply chain and in turn make it better respond to the PoS data. In the case of the up-scale retail centers with a limited backroom space at their end, pool distribution would help stores reduce the inventory levels and increase their ability to frequently refresh shelves. Manufacturers can consolidate their inbound orders for supply of components/parts from different suppliers and transport larger outbound shipments. Thus, flow-through distribution, in addition to realizing economies of scale in transportation, also supplements JIT/VMI (vendor-managed inventory) processes and eliminates the need for warehousing inventory.

iv. **ABC**: It entails tracing the overhead and direct costs back to processes related to specific products and services. Earlier, it was much easier and realistic to allocate overheads and direct costs based on the number of hours required to produce a product, but now since the direct labor costs have reduced to just 10% or so of the total costs, ABC had to be used. In fact, ABC is a part of ABM (Activity-Based Management), which comprises continuous improvement and business process analysis in addition to ABC. ABC also aims to identify the cost drivers also like the number of orders, length of set-ups, specifications, engineering changes, and the follow-up and expediting required. So, ABC can be used to identify and implement cost-saving opportunities on the supplier side. The purchasing and supply manager must encourage the supplier in understanding as to how the supplier estimates and comprehends overheads to the product being purchased. As the overheads become a greater proposition of product cost and as purchasing and supply managers strive to always reduce the cost of acquisition of materials, accuracy in estimating and applying overheads can affect the final costs and thus the pricing.

v. Category Management: It is aimed at maximizing the potential of a product category by focusing on variety, merchandising, stocking, pricing, introducing the new products, differentiating, or resupplying products in a certain manner. Category management can increase retail sales per square foot by advising companies to selectively focus on categories that offer the best return on the investment made in any resource; the key being the inventory, others could be space and personnel. The activities managed include developing a strategic category plan by use of PoS data. It is then ensured to monitor space allocations and inventory replenishments and measure each category's performance at retail level on a regular basis for continuous improvement. In the current environment, it may also involve joint working with suppliers to produce products at a target cost driven by the price (the Japanese approach) rather than a price driven by costs (the American approach). Companies like SAP offer category management solutions that help the company to plan and manage retail goods on a seasonal sales pattern basis. The final targets for each category are defined in terms of retail space allocations, number of orders placed/frequency of replenishment, product hierarchy design, and product innovations and differentiation. Category managers then aim to accomplish these targets. Overall, category management would help in multi-channel retailing, inventory management, stores, and pricing management to meet profit margins in a product category. Also, the cost of exposure of a category (of being caught with a stock-out at retail level) is estimated before planning a replenishment pattern for a category.

vi. EDI: It is needed to transmit and manage orders and their delivery, invoices, and payments both within the company and outside with its business partners. In fact, EDI forms the basis of implementation of ECR.

2.4 MECHANICS OF ECR

ECR could transform the working of the entire FMCG industry for starters. With ECR, now companies can cross-utilize resources and improve efficiencies further. That means competitors can now share trucks and warehouses, use common business language, and jointly tackle problems like excessive inventory and stock-outs.

ECR's objectives are simple – fulfill consumer demand better and faster and at a lower cost. ECR, then becomes like a lubricant applied all along the FMCG supply chain. The two operating doctrines that guide ECR efforts are the focus on consumers, their demands and expectations, and working together within the organization internally and with the business partners externally to overcome the barriers. The three focus areas are demand, supply, and enabling technologies as shown in Figure 2.2.

The demand side of ECR includes all the flow of activities associated with estimating and managing the demand for products and services and their specifications by the customers to serve shoppers more effectively, their satisfaction, and contributing

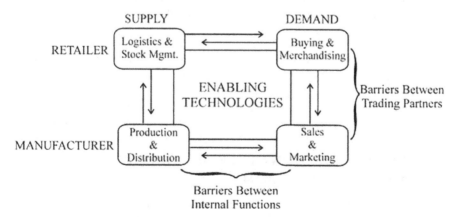

FIGURE 2.2 Structure of network and flows in an ECR system. (www.gs1india.org/about/ECR.htm as retrieved from 14 October 2008.)

to increase in sales and market share for trading partners. The typical priorities with ECR management could lie in the following:

a. Running a few demand pilots just to know the specific needs and capabilities of collaborating with each other for demand planning.
b. Rationalizing the assortment at retail level, i.e., phasing out low-performing SKUs and improving the location of a few of them depending on retail business goals, e.g., stores productivity in terms of resources like space and inventory (turnover).
c. Optimizing the promotions in terms of the promotion objectives jointly set with trading partners and their execution in terms of internal and external communications, timely distribution, and placement of campaigns. The purpose is to reduce the response time of customers.
d. Jointly working for new product introductions by screening products against certain benchmark factors early in design and development. It may involve joint working on product launch/test market reports. The purpose obviously is to reduce product design and features' complexities and reduce the time and cost to design, develop, produce, and market it ahead of competitors.

The supply side of ECR refers to production and distribution planning, scheduling and execution by the manufacturer, and logistics and stock management by the retailer. This must be done by overcoming the barriers between the trading partners, e.g., by sharing business information on a real-time basis.

Enabling technologies are required in ECR networks for sharing of timely and accurate information between trading partners. It mainly refers to developing data processing and management capabilities of the ECR system. The priorities may involve the following:

a. Developing a standardized product identification in terms of bar codes or EAN or EPC and format for EDI. It may involve joint working both internally with other departments and externally between trading partners.
b. Re-engineering the ECR processes to reduce their cycle times and bring about simplification.

For the intermediaries, ECR would enable a fundamentally different way of functioning – from push-based, where the distributor would dump the supplies on retailer irrespective of the demand or order level or sales pattern to demand or *pull*-based, where the trading partner picks products based on the actual demand or number of orders only. Distributor would not supply unless there is a demand/order trigger communicated to it.

2.5 SPECIAL APPLICATION OF ECR IN PRODUCT CATEGORY MANAGEMENT

In the ECR concept, the assortment is seen as a body comprising separate product categories according to consumers' needs and psychological aspects of product purchase. Every retail trade enterprise works out its product categories based on consumers' needs (Hofstetter, 2006).

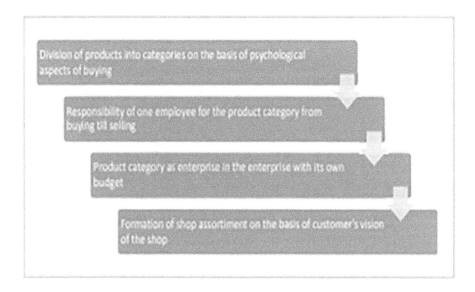

FIGURE 2.3 Product category management process. (Seifert, 2006.)

Consumer research and segmentation is thus a prerequisite for ECR. What product category does the customer fit-in, and how does she view the particular store in context of the category is an important part of category management. The category management process is given by Seifert (2006) (Figure 2.3).

It is recommended to designate one executive as a product category manager. The responsibility of the product category manager would encompass the following:

a. Sharing of information with the suppliers and logistic providers
b. Assortment planning
c. Regular product replenishment on the shelves phasing out nonliquid merchandise or SKUs
d. Define zone of responsibility for sales performance in a category
e. Pricing a product category
f. Treating each category as an enterprise within the company and thus assigning operative and financial goals for the same
g. Preparing a detailed business plan for the category and getting a budget allocated for the category from top management
h. PoS material placement and sales promotion

Product category management thus increases the avenues of profit from sales and competitiveness of the enterprise. It is therefore imperative to do an assessment of the competitiveness and positioning, preparing the policies for assortment, pricing, cooperation with suppliers, and dedicating a product category manager with a well-defined job description and zone of responsibility for performance in a category.

According to a study by Zvirgzdina, Linina, and Vevere (2015) in Latvia, a lack of understanding of product category management and not recognizing its role in

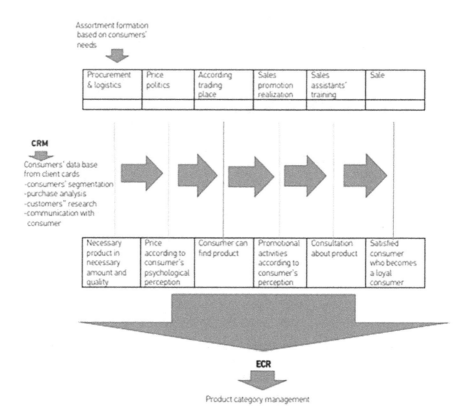

FIGURE 2.4 ECR model in retail trade enterprises. (Zvirgzdina et al., 2015.)

strategic planning can well be an obstacle in implementation of ECR. Zvirgzdina et al. (2015) also offered a model (Figure 2.4) for the introduction of the ECR in the retail trade enterprises. As is seen in the model, it starts from assortment formation based on consumer needs, using CRM (customer relationship management) data, e.g., spend analysis followed by developing an ECR as derived from variant specification, promotion, pricing, availability, extent of customer loyalty, and other consumer needs and expectations.

2.6 UNIVERSAL PRODUCT CODE (UPC)

The purpose of UPC is to identify unique product features and to ease out the checkout process at the retail stores. UPCs help in *track and trace* and end-to-end visibility of the products throughout the supply chain. These codes help in identification of each variant, hence SKU (stock keeping unit) and better category management. As per GS1, a global standards organization, over 2 million companies globally use GS1 barcodes that are being scanned 6 billion times a day.

2.6.1 FEATURES OF UPC BARCODE

  i. In order to use a UPC code, the manufacturer of a product must submit an application to GS1 and become a member. Once approved, GS1 will assign a unique six-digit code that identifies only that manufacturer. These six digits form the first part of the 12-digit UPC. The next five digits are the item number itself.
 ii. Many consumer products have several variations, based on, for example, size, flavor, or color. Each variety requires its own item number. So, a box of 96 one-inch screws has a different item number than a box of 96 two-inch screws or a box of 24 one-inch screws.
iii. The last digit is the check digit. It is the product of certain unique calculations to confirm to the checkout scanner that the UPC is valid. If the check digit code is incorrect, the UPC won't scan properly.
 iv. UPC barcode consists of a scan-able strip of black bars and white spaces, above which contains a sequence of 12 numerical digits as mentioned. No letter characters or other content of any kind may appear on a standard UPC barcode. The digits and bars maintain one-to-one correspondence, i.e., there is only one way to represent each 12-digit number visually, and only one way to represent each visual barcode numerically.
  v. Once a product is discontinued from circulation, the number is phased out so that it will not be used again for another product.

UPCs provide higher online visibility to differentiated product variants, thus helping the company gain more recognition in the competitive marketplace. A Global Trade Item Number (GTIN can be used by a company to uniquely identify all of its trade items. According to GS1 standards organization, on its website, www.gs1.org, trade items are defined as products or services that are priced, ordered, or invoiced at any point in the supply chain. GTINs are extensively used in health care, fresh food, retail, and rail industry. GS1 barcodes and GTINs are used with most online and traditional retailers including Amazon, eBay, Alibaba, Google, Carrefour, Tesco, and Walmart.

An Electronic Product Code™ (EPC) is a unique identifier assigned to physical objects, unit loads or pallets, locations, or other identifiable entities existing in a supply chain. There is interoperability between GTIN, UPC bar code or EPC and RFID (Radio-frequency Identification). The RFID-based EPC tags can be used for EAS (Electronic Articles Surveillance) functionality for deterring and detection of loss and pilferage of consumer goods in the retail industry particularly at PoE (Point of Entry), Point of receiving, and Point of sale (PoS) in the FMCG retail industry.

2.7 OVERVIEW OF GDSN

GDSN is a global network that facilitates the synchronization of item information between retailers and suppliers by using a single global registry. It is a single PoE into the network for suppliers and retailers through selected data pools. Once connected,

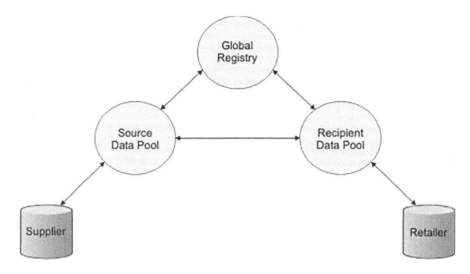

FIGURE 2.5 GDSN synchronization process. (https://www.ibm.com/support/knowledge-center/SSWSR9_11.6.0/com.ibm.pim.ovr.doc/pim_con_gds.html, Date accessed: 8 February 2021.)

the trading partners can subscribe to or publish all of their item information in a single and consistent process for all other trading partners that are connected to the GDSN. The item information could be about price and other product attributes and specifications. Figure 2.5 shows the flow of product data between trading partners, data pools, and the global registry within the GDSN.

The GDSN synchronization process as given in Figure 2.5 involves the following steps:

i. The supplier submits information about the product to the source data pool.
ii. The source data pool registers the item in the global registry, which then helps the GDSN community to locate data sources and manage ongoing synchronization relationships between trading partners.
iii. Subsequent to the registration of the item at the global registry, the supplier publishes the item in the source data pool.
iv. The retailer subscribes to the item by sending the subscription message to the recipient data pool.
v. The recipient data pool along with the subscription details requests information about the item from the source data pool through the global registry.
vi. Based on the retailer's subscription information, the source data pool synchronizes with the recipient data pool to share the item information which includes price, attributes, specifications, and other product description.
vii. The recipient data pool forwards the item information to the retailer after retrieving information from the source data pool.
viii. The retailer confirms the recipient data pool if the item is approved or rejected following which the recipient data pool sends an item confirmation to the source data pool. A Data Recipient is any business entity – be it a medical

The Role of UPC and GDSN in ECR

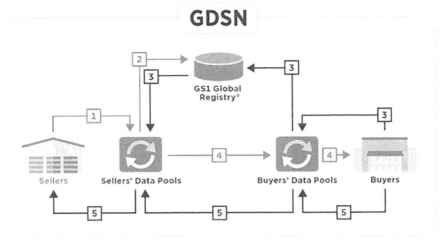

FIGURE 2.6 The flow of data in GDSN. (Source: https://www.gs1.org/services/gdsn/how-gdsn-works, Date accessed: 6 February 2021.)

organization, retailer, distributor, consumers' group, or manufacturer – that subscribes to Trade Item information.

ix. The source data pool finally forwards the item confirmation to the supplier.

The data flow through GDSN are depicted in Figure 2.6.

2.7.1 Product Information Sharing through GDSN and Impact on ECR

GDSN will improve the efficiency of getting globally sourced products onto the retailing shelves, thereby satisfying the consumers by

- Enabling online visibility of product data and product availability through GLN (Global Location Number) accurately across the world.
- Enhancing credibility of retailers by allowing the manufacturer to track the products globally.
- Entertaining regulatory requests for traceability and product recalls using GTIN and GLN.
- Incorporating use of product variants' reliable information related to price and attributes through GTIN.
- Enabling flexibility to quickly respond to segmented market needs and thus ensuring an ECR suited to consumer needs and expectations ultimately leading to higher consumer service levels and thereby business excellence and growth.

2.8 ECR IN INDIA

ECR India got registered in July 2000 as a nonprofit 'Association of Persons (AoP)'. EAN India is an affiliate of GS 1 (formerly EAN International), Brussels, which has 101 EAN organizations representing 155 countries worldwide. EAN India, now GS 1 India, is promoted by the Ministry of Commerce, Govt. of India and was registered in 1996 as a society for promoting the use of GS1 numbering system for unique identification of a product/SKU and location in line with international best practices. Its board of management has members like APEDA, BIS, ASSOCHAM, FICCI, CII, IMC, FIEO, and IIP. Over 1 million companies use GS1 standards in the world. Initial member companies of ECR India have been Johnson & Johnson, Nestle, Food World, PwC, and EAN (European Article Number) India, now GS1 India. They found that there were a lot of areas where collaboration could lead to lower costs and better supply chain management. The number of participants steadily increased. ECR India includes HUL, Nestle, Colgate, Glaxo SmithKline Beecham, P & G, Johnson & Johnson, Godrej Consumer Products, ITC, Cadbury India, Marico, PwC, Wipro Consumer Care & Lighting, Agro Tech Foods Ltd, Cavin Kare, Food World, EAN India, Kodak India, Subhiksha Trading Systems, Henkel Spic, and the transportation giants like TCI, South Eastern Roadways, and Container Corporation of India.

2.8.1 An Example of ECR Initiative in India

To understand as to how companies are implementing ECR initiatives in India, an example has been analyzed by Sharma (2013) at the arrangement between Nestle and J&J. These companies have worked it out at Delhi, a transit point for products leaving Nestle's manufacturing plant in Moga, Punjab. A common transporter Transport Corporation of India (TCI)'s trucks leaving Delhi with goods meant for Nestle's distribution in Mumbai come back loaded with J & J labels for the Delhi market. Such tie ups for back haul facilities have resulted in 5%–10% savings in the transport costs, besides cutting down transit time. If warehouses are consolidated across companies, then even deliveries could be shared, leading to further savings. These 'backhaul' arrangements normally consist of a shared transporter.

HUL, Johnson & Johnson, and Nestle are further working toward drawing up a common format for data exchange between 200-odd manufacturers, 30,000-odd distributors, and thousands of clearing and forwarding agents (C & FAs), wholesalers and 12 million-odd retailers all of whom together make up India's huge and complex supply chain.

2.8.2 Workgroups in ECR India

The following are the four workgroups set up by the ECR India:

   i. Out-of-Stock Workgroup
  ii. Logistics Workgroup
 iii. Dataflow Standards Workgroup
  iv. FMCG Policy Group for Organized Retail.

2.8.3 THE FUTURE OF ECR IN INDIA

The way ahead for ECR India is to get some substantial benefits from the projects under progress and use these as a selling point to get more companies on board. This would help in having more broad-based benefits from the projects in future, which would help in realizing the mission of ECR India of

> removing unnecessary costs from the supply chain and making the sector, as a whole, more responsive to consumer demand.

As one Colgate Palmolive spokesperson put it in one of the workgroup meetings, 'Our ECR initiative in India is part of our global strategy to improve customer service and achieve margin improvements. ECR has a great potential to make it big in India'.

India needs more corporate members of the ECR movement to participate, and it needs to be implemented across different sectors particularly in services for enhanced consumer response.

2.9 ECR IN WORLD

There are big changes taking place in the ECR world. In Baltic region, there is the ECR Baltic group, which has done pioneering work in ECR in Baltic countries of the world. ECR Baltic is an ECR initiative in Estonia, Latvia, and Lithuania. It is a collaborative retailer-manufacturer platform with a mission 'to fulfill consumer wishes better, faster, and at less cost'. More information is available on www.ecr-baltic.org.

It has made work groups as follows:

a. Competition law and antitrust motion
b. Digital (connected business information) committee
c. European Action Projects
d. Fair Trading Practice Initiative
e. Education and Training.

In an interview quoted by Zvirgzdina et al. (2015), Director of ECR Baltic, Edgars Pentjuss, cited the retailers' failure to treat the suppliers as partners and to share information with them, the imperfection in supplies, lack in use of modern technology, unavailability of specialists and resistance of the personnel as bottlenecks in implementation of ECR.

Consumer Goods Forum is another organized initiative for ECR in Europe. Companies like P&G, L'Oreal, and Carrefour have gone a big way in implementing ECR. IBM has been a consultant in these projects. IBM global business services published in 2011 a report on using GS1 standards for efficient supply chain through GDSN. They developed KPIs in terms of business measures and implementation measures. Business measures include unit fill rate, PoS stock-outs, invoice accuracy, lead time, supplier service level, and number of days of finished goods inventory cover with suppliers. Implementation measures include consumers allocated GTIN, number of locations assigned GLN, percent of sales achieved from trading partners with synchronized master data shared by GDSN, percent pallets or shipping units assigned SSCC-Serial Shipping Container Code, number of orders transacted using EDI, and number of invoices transacted using EDI. In the USA, there is a good organization of the grocery industry, e.g., FMI and Grocery Manufacturers Association. Then, the set of guidelines under initiative of CPFR started by VICS (Voluntary Interindustry Commerce Solutions) is worth consideration.

2.10 CONCLUSION

ECR enables the supply chain members to work together to fulfill consumer needs better, faster, and at reduced costs. Cost-saving, supply efficiency, profit increase, and better consumer service levels can be accomplished from the implementation of the GDSN, Product Category Management, CRP, barcoding/unique product coding (UPC), EDI, CAO, flow-through distribution/cross-docking, and ABC among other strategies of the ECR. Most of the data between supply chain buyers and sellers in the FMCG industry flow through GDSN facilitated by the global GS1 organization. Various ECR work groups established in the USA, Europe, Baltic region, and India can go a long way in arriving at solutions aimed at enhancing customer value.

ECR should be designed to configure and integrate the supply chain from 'push' structure to the 'pull' structure where the merchandise is controlled by the actual consumer demand at direct PoS rather than pushing on consumers schedule-based finished goods stocks. Hence, nothing is more important than sharing of actual sales-based information and data exchange among supply chain members through networks like GDSN and product coding and traceability systems like GTIN-embedded EPC/RFID tagging and management practices like category management using a host of consumer centric initiatives for business excellence and growth.

REFERENCES

Arango, M. (1999). 'Benefits of Inter Firm Coordination in Food Industry Supply Chains', *Journal of Business Logistics*, Vol. 26 no. 7, pp. 20–22.

Brockman, M. and Morgan, R. (1999). 'The Evolution of Managerial Innovations in Distribution: What Prospects for ECR?', *International Journal of Retail & Distribution Management*, Vol. 12 no. 5, pp. 88–92.

Brown, T.A. (2001). 'ECR and Grocery Retailing: An Exploratory Financial Statement Analysis', *Journal of Business Logistics*, Vol. 22 no. 2, pp. 77–90.

Coopers, L, (1998). *The Grocery Industry Supply Chain Committee: 1998 Tracking Study*, Grocery Manufacturer of Australia Ltd, Sydney.

Ellram, L. and Weber, M. (1989). 'Retail Logistics', *International Journal of Physical Distribution & Materials Management*, Vol. 19 no.12, pp. 29–39.

Greenbaum, J. (1997). 'Efficient Consumer Response: How Software Is Remaking the Consumer Packaged Goods Industry', *Software Magazine*, Vol. 17 no.1, pp. 38–48.

Griffin, J.M. (1998). 'Effect of a Mass Merchandiser on Traditional Food Retailers', *Journal of Food Distribution Research*, Vol. 29 no. 4, pp. 123–129.

Hall, E. and Hall, M. (1990). 'Understanding Cultural Differences, Intercultural Press. Strategy for Grocery Businesses', *International Journal of Service Industry Management*, Vol. 11 no.4, pp. 365–370.

Hofstetter, J.S. (2006). 'Assessing the Contribution of ECR', *ECR Journal*, Vol. 6 no. 1, pp. 20–29.

Holms, J. (1998). 'Implementing Vendor-Managed Inventory the Efficient Way: A Case Study of Partnership in the Supply Chain', *Production and Inventory Management Journal*, Vol. 39 no. 3, pp. 1–6.

King, R.P. (1996). 'Reengineering the Food Supply Chain: The ECR Initiative in the Grocery Industry', *American Journal of Agricultural Economics*, Vol. 8 no. 78, pp. 1181–1186.

Kurnia, S. and Johnston, R.B. (2001). 'Adoption of Efficient Consumer Response: The Issue of Mutuality', *Supply Chain Management: An International Journal*, Vol. 6 no. 5, pp. 230–241.

Kurt Salmon Associates (1993). *Efficient Consumer Response—Enhancing Consumer Value in the Grocery Industry*, Food Marketing Institute, Washington, DC.

Perona, M. (2004). 'A New Framework for Supply Chain Management Conceptual Model and Empirical Test', *International Journal of Operations*, Vol. 22 no. 14, pp. 123–134.

Seifert, D. (2006). 'Efficient Consumer Response', In: Zerres, C., Zerres, M.P. (eds.) *Handbook Marketing-Controlling*. Springer, Berlin, Heidelberg.

Sharma, S. (2013). *Supply Chain Management; Concepts, Practices and Implementation*, Oxford University Press, New Delhi.

Zvirgzdina, R., Linina, I. and Vevere, V. (2015). 'Efficient Consumer Response (ECR) Principles and Their Application in Retail Trade', *European Integration Studies*, Vol. 9, pp. 257–264.

3 Delivering Superior Customer Experience through New-Age Technologies

Vaishali Kaushal and Rajan Yadav
Delhi Technological University

CONTENTS

3.1 Introduction 47
3.2 Customer Experience (CX)...48
3.3 Augmented Reality (AR)..49
    3.3.1 Impact of AR on CX...49
    3.3.2 Industry Examples of Augmented Reality Enhancing CX................50
3.4 Artificial Intelligence (AI)..52
    3.4.1 AI in the Domain of CX..53
    3.4.2 Industry Examples 53
    3.4.3 Issues with AI 54
3.5 Chatbots..54
    3.5.1 How Chatbots Deliver a Superior CX 55
    3.5.2 Issues with Chatbots 56
3.6 Conclusion 57
References..57

3.1 INTRODUCTION

Customer experience (CX) has lately grabbed the attention of all marketers. Today, more than pricing, a customer is affected by the holistic experience of making a purchase. CX comprises a consumer's cognitive, affective, behavioral, sensory, and social response to an organization's offering during the entire purchase journey.

    Digital transformation has revolutionized the way companies work as it helps companies adapt to new market realities and change business models. These new-age technologies carry the full potential to augment the customer experience across all touchpoints. The digital transformation has brought a paradigm shift in customer expectations. A customer is constantly connected with organizations via technology and aware of what all can be done with technology.

DOI: 10.1201/9781003140474-3

Digital technology bestows the power to augment any part of the customer journey, including active or passive touchpoints. Active touchpoints comprise person-to-person or person-to-machine communication, whereas passive touchpoints initiate communication in the background. Features like biometrics and face recognition allow examining consumer behavior, movement patterns, and interaction with digital touchpoints. Also, recommendations based on browsing patterns or purchase history personalize a customer's feed, delivering an enhanced CX. For example, personalized content and special offers make a customer feel important and delivers an enhanced experience. Near-field communication technology has replaced traditional payment in retail stores. Inside a store, mobile technology has the power to augment the service scape like virtual reality. New technologies like AR provides the customer with a smooth omnichannel experience by integrating channels and narrowing the gap at various online and offline touchpoints. So, all these technologies make the CX richer every day by enhancing experiences around different digital touchpoints.

3.2 CUSTOMER EXPERIENCE (CX)

Companies have started to recognize the importance of delivering an enhanced CX. A customer's experience can be defined and gauged in multiple ways. Jain, Aagja, and Bagdare (2017) defined it as 'the aggregate of feelings, perceptions, and attitudes formed during the entire process of decision-making and the consumption chain involving an integrated series of interaction with people, objects, processes, and environment, leading to cognitive, emotional, sensorial, and behavioral responses'. customer experience encompasses all the interactions of a customer with its touchpoints during the customer journey. Hence, it creates a long-lasting impression which helps the customer form a repurchase intention from the same brand.

Retailers need to concentrate on the online customer experience to perform successfully in today's competitive and changing environment (Grewal et al., 2009). On a digital platform, customer experience is formed by the various interactions and quality of the experience, from navigating to making the purchase. Moreover, customer experience is complex and contains cognitive, social, affective, and physical elements (Verhoef et al., 2009). Several word combinations for the online experience exist in the literature, such as 'online CX' (Rose et al., 2012), 'website experience' (Kim et al., 2009), and 'online purchase experience' (Holloway et al., 2005). Rose et al. (2012) define online customer experience as a psychological state manifested as a subjective response to the website. The customer is involved in the cognitive and affective process of incoming information from the online platform, and the outcome creates an impression in the consumer's memory. Individually, a better and more customized experience leads to improved online shopping satisfaction (Kaushal & Yadav, 2020). The technology acceptance model given by Davis (1989) mentions that usefulness, effectiveness, and ease of use help building a positive perception toward new technologies. Hence, the customers will accept AI/AR if they believe it is easy to use and brings more convenience (Trivedi, 2019). However, the main questions that arise are the following:

1. What are these new-age technologies, and how do they augment CX?
2. What are the features and benefits of these technologies?

3. What are the issues with these technologies like AR, AI, and chatbots?
4. Which industries/organizations have been using these technologies, and what have they achieved?

3.3 AUGMENTED REALITY (AR)

AR is a distinctive technology that sparks imagination and carries the ability to engross a customer's cognition and emotion by creativity and playfulness (Heller et al., 2019; Rauschnabel et al., 2019). AR helps integrate physical and virtual objects and augment holistic CX.

Huang et al. (2015) highlight AR as a technology to enhance CX and customer satisfaction. Beck and Crié (2018) attested the AR's impact on purchase intention. AR delivers an interactive and consistent experience that facilitates users to get acquainted with audio-visual elements. One of the main aspects of this technology is that it encourages interaction. AR is 'a technology which allows computer-generated virtual imagery to exactly overlay physical objects in real time' (Zhou et al., 2008). AR-based technologies are meant to be interactive and user-friendly. A customer should be able to click on a hyperlink to read product details and feedback, choose colors from the palette, and experience an overall customized experience. Poushneh (2018) mentioned that retail websites have increased interactive characteristics.

3.3.1 IMPACT OF AR ON CX

Optimal CX is created with a confluence of virtual physical touchpoints (Javornik, 2016) which expands customers' horizons to more dynamic experiences (Ostrom et al., 2015). And with the rise of new-age technologies, we should be able to improve the customer journey. Azuma (1997) believes AR to be a blend of the actual and virtual world yet in real time. AR bestows the power to create an entirely artificial environment, thereby supplementing reality with digital content (Flavián et al., 2019). AR devices can be stationary, movable, or wearable. Google Glass, one of the first examples of an AR smart glass, provided an enhanced virtual experience to the wearer. (Han, Tom Dieck, Jung, 2019) Microsoft has come up with AR smart glasses (ARSG) wherein the device is believed to incorporate virtual information into the wearer's visual field (Kalantari & Rauschnabel, 2018).

Preexperience: Imagine a furniture brand-consumer able to foresee how a particular corner in their house would look with new décor or some furniture. A theme park could include an AR experience of a rollercoaster in a shopping mall. Another example could consist of playing in a virtual world with historical avatars. A different idea could be scanning a tourism brochure with an AR-enabled mobile app to have a taste of how the travel experience would be.

During Experience: An example of during-experience stage of AR would be using AR glasses to view digital content. For instance, tourists use AR and VR devices to experience the process of making wine while on wine-tasting tours. AR is also used by drivers to look at real-time GPS information on their windshields while driving.

Postexperience: Consumers could receive prompt assistance on how to repair a machine which can be assisted via a video call with a chatbot. For an art gallery, visitors could be invited to rate the paintings and gallery services through AR applications.

FIGURE 3.1 AR-enabled mobile application. (Created by Author: Vaishali Kaushal)

Scholz and Smith (2016) empirically validated whether the use of AR and VR maximizes customer engagement in their research. Rese et al. (2017) focused on variables like ease of use, aesthetics, and usefulness of AR applications. Consumers through their gestures can manipulate virtual products and record various behaviors that permit consumers to review and recheck various effects. These augmented realities play an unprecedented role that can not only create a positive effect but also create an online brand experience. Many studies focus on how AR technology can be applied in various other fields (Beck & Crié, 2018). AR also influences the sensory perceptions of customers related to their purchase decision. Figures 3.1 and 3.2 depict how AR based apps automatically scales furniture as per the room dimensions and helps visualize the product thereby leading to better decision making.

3.3.2 Industry Examples of Augmented Reality Enhancing CX

- Ikea has developed the 'Ikea Place' mobile application using Apple's ArKit technology. Consumers can select a virtual picture of any furniture and 'set it up' in their house by directing the camera to the desired areas. Smartphone

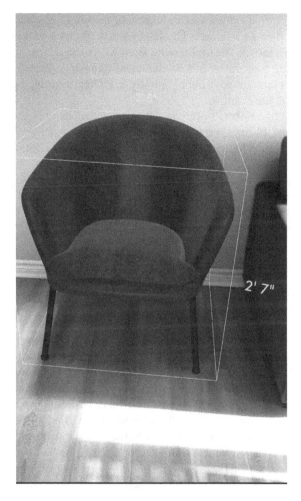

FIGURE 3.2 Furniture with dimensions using AR. (Created by Author: Vaishali Kaushal)

cameras scan consumers' homes and superimpose the selected set of furniture displaying a virtual view of how it would look. Thus, consumers can 'try out' different pieces and select the best one for their house. This also drastically reduces postpurchase dissonance.
- Various brands such as Adidas, American Apparel, Lacoste, Loreal, and Sephora already offer AR technology on their mobile website.
- AR's popularity boom commenced with the video game Pokemon Go that garnered global attention (Rauschnabel et al., 2017).
- Footwear brands like Nike and Gucci provide AR features to try out different shoe designs and might consider including social sharing so that consumers can instantly acquire feedback from friends (Hilken et al., 2019).
- The use of AR technology by 'body brands' like clothing and cosmetics like De Beers and eBay enhances brand attachment (Yim et al., 2017). Customers

can try on various clothes and shoes with the click of a button from the comfort of their own homes. The sales of such products will only go on an upward trajectory as AR proliferates. With AR-enabled virtual stores, a firm can stimulate consumers' purchasing habits (Kumar & Paul, 2018). AR can be used as a tool for 'envisioning' products; L'Oréal's 'trying before buying' is an appropriate example.
- Restaurants like KabaQ's offer an AR-enhanced menu to create playfulness around their offerings, provide a novel gastronomy experience, and enable visitors to select the 'right' dish by providing virtual photos of the food on offer.
- It could translate into a greater visit intent to give tourists a 'try before you buy' experience of their selected destination (Marasco et al., 2018). The technology's convincing capacity for tourism lies in its capacity to overcome the intangible before visiting a destination.

3.4 ARTIFICIAL INTELLIGENCE (AI)

Virtual assistants, driverless cars, navigation software, wearable products, product suggestions (Netflix, Amazon.com), and face recognition software (Apple, Facebook, mobile lock screens) have revolutionized the status of artificial intelligence (AI) and made it ubiquitous (Holder et al., 2016). The 21st century was transformed by faster computers, big data, and deep learning, which triggered the revival of interest in AI (Buchanan, 2005). Roe (2017) discussed AI's capabilities to improve salesperson-customer relationships. AI-based chatbots can provide real-time answers to customers' queries and with accuracy. Colon (2018) reported a study wherein customer retention was improved by 80% when the sales team was using AI. Self-service technologies (SSTs) are a blessing to customers as they allow them to browse items and place orders as well as track them. Alibaba.com, a B2B trading firm, banked upon value addition by these SSTs, thereby reducing informational asymmetries. These technologies are altering the service interface from human-driven to technology-dominant (Larivière et al., 2017).

Brands have created a powerful brand image in the consumer's mind and have strong perceptions of trust and privacy (Brill et al., 2019). Cohn and Wolfe (2017) mentioned that 75% of consumers are willing to share their information with the brands they confide in. The same study identified firms like Apple, Google, Microsoft, Amazon, and Paypal to be the most trustworthy brands, leaving behind Facebook due to past privacy issues. This elucidates the fact that the size of a firm does not guarantee consumer trust. Leung and Loo (2020) mention the smart service networks being used for the postpandemic dining experience that is helpful to both customers and restaurant operators – from information collection to personalization and from sharing sensory information to interacting with customers to delivering meals by automated robotic technologies. Such networks require a large amount of data interchange and interconnected platforms to reduce human intervention (Buhalis & Leung, 2018). Lui and Lamb (2018) discussed how robots and chatbots have made the whole banking process customer-centric and more personalized.

3.4.1 AI IN THE DOMAIN OF CX

AI is an overall term for computers with human-like capabilities of hearing, seeing, reasoning, and learning (Güngör, 2020). Today, AI has become an indispensable part of our lives.

Consumers are pressed for time and want superior customer service. Rather than costs, the CX is given weightage. SST is gaining traction as customers like the DIY compact approach backed by bots have replaced customer agents to save time and energy across channels and afford seamless interactions. AI combined with real-time data personalizes experiences. AI enables brands to enhance their content delivery, create effective customer segments, and deliver data visualization. It helps shopping websites make visual recommendations based on what the customer already sees on their screen. AI helps monitor consumer interaction through browsing data, email open history, click-through rates, and various other actions (Riener et al., 2019). It can handle a multitude of customer queries resulting in a more complete and engaging experience. For firms with a large customer base, AI has become even more critical.

AI is used for variable pricing in travel and hospitality, fraud detection, and automatic trading tools. Some other trends include low-cost storage, faster processors, connected devices, machine learning innovations, and cloud AI.

3.4.2 INDUSTRY EXAMPLES

- 'Nadine', a human-like robot, was brought on as a receptionist in Singapore's Nanyang Technological University (NTU) in 2018. She was designed to fulfill administrative tasks and focus on staffing. She can engage with her surroundings, demonstrate emotions, and work for long periods (Nanyang Technological University Singapore, 2019). 'Nadine' could be a starting point for the travel and hospitality industry.
- Makadia (2018) mentions some more examples like 'Connie' in Hilton Hotels & Resorts that can perform all tasks which stir pure delight in the hospitality industry. For example, Connie can greet guests in different languages, recommend attractions, and assist in smooth check-in (Solomon, 2016).
- Wynn Casino in Las Vegas is a classic example of enhancing CX as it has integrated Amazon Echo to digitalize hotel rooms, offering voice control for TV, smart lighting system, and AC temperature so that guests feel valued.
- Banks can use AI for some customers who do not pay their bills on time, putting their credit score at risk.
- KFC's collaboration with China's Google Baidu has led to facial recognition at a different level where it can predict what the customer will want to eat. This system considers the age, gender, mood of the customer, and previous ordering history. It is designed to process a customer's characteristics and offers a curated menu before the customer makes a selection.

3.4.3 Issues with AI

Consumer privacy and security are some of the serious issues plaguing AI. For example, banks and other financial institutions hold sensitive data – giving robots unrestricted access to this can have devastating consequences. So, data management should involve proper data protection to prevent it from being misused or hacked.

Also, AI can pose challenges related to bias and discrimination, which need to be regulated. Innovation around consumer data protection is the need of the hour. So, a high level of AI needs to be used so that it interprets human emotions well and can provide personalized solutions by delivering genuinely hospitable service. AI has the potential to improve organizational function and achieve higher customer satisfaction (Osawa et al., 2017). An example of failed AI is of three restaurants where robotic waiters were 'fired' within 1 year of operation due to inept service and frequent technical difficulties despite the huge investment. Another example would be a travel company, where one-third of the workers were replaced by machines in an amusement park. Later, management had to make changes in their personnel strategy as the robots posed even more challenges (Hertzfeld, 2019). Developing rules and principles such as addressing robots by names, defining values, and integrating them as a new system member can reduce the adverse impacts of technology on socialization. Humans would still have an irreplaceable role as robots cannot provide human judgment on a real-time basis. Another drawback is that robots miss out on the empathy factor and can never understand the feelings of a human; thus, robots can help in decision-making only to an extent.

3.5 CHATBOTS

Live chat services denote web-based functions, which facilitate users to interact in real time with customer service agents (Elmorshidy, 2013). The AI technology uses human language for coding and decoding to understand interactions and respond accordingly (Hill et al., 2015).

Chatbots are designed with an underlying objective to immediately respond to customers' queries. This coupled with the greater convenience has led to this technology adoption by consumers. The chatbot is a computer programmed to formulate conversation and frame reverts in the form of written texts or voice methods (Miner et al., 2020). AI-based chatbots are described as follows – Chatbot is an acronym for a chat robot that aims to establish communication with humans with the help of AI technology (Jones et al., 2018). Chatbots as per Schlicht (2016) are 'services powered by rules and sometimes artificial intelligence that you interact with via a chat interface'. For instance, consumers may write in a question or statement and will receive a reply in a conversational style. Customers 'text' their queries as if they were chatting with another human, and the chatbot replies similarly. Chatbots carry the potential to serve diverse online customers and deliver customized support, site guides, virtual support, etc. Dual chatbot attributes, i.e., of a social actor and a new technology are being talked about. McLean et al. (2019) discusses the functions of live chat services which enhance the CX, help businesses respond to online customer questions, augment online social interactions, and personalize shopping experiences.

The chatbot interface is via smart devices which do not require any prior downloads and can be used immediately, posing ample competitive advantages. While clients search for results such as digital guidance, content, and solving queries/issues, firms also seek solutions for elements of the industry. As per Gartner, 25% of customer service operations have integrated chatbots across various engagement channels by 2020. Chatbots help companies provide 24×7 support, improve engagements, and save valuable time. AI-powered chatbots have the capabilities to make lives easier for customers as well as service providers. Most of the queries which companies get are routine ones (mostly FAQs), but customers today need an ear to listen to their questions and grievances. Hence, chatbots enable human resources to focus on complex tasks that require human intervention. As a result, AI-enabled chatbots help boost the productivity and efficiency of a company. One example of the same is that Edward by Edwardian Hotels can communicate in natural, conversational language to guide tourists throughout the entire travel journey. Guests communicate with Edward as if they're talking to a fellow human. Edward manages 69% of the queries of the guests and the re-assignment of staff from repetitive tasks to more important queries (Oram, 2019). As the marketing adage goes, 'The customer is king'. But when customers are expected to wait for hours for a response from customer support or a reply to their emails, they usually give up and go. Consumers sometimes also have to scroll through numerous pages of an organization's website to locate the contact details of the firm. Chatbots provide the power to automate support functions doing the job of virtual support, customer service representatives, and sale agents (Moriuchi et al., 2020).

3.5.1 How Chatbots Deliver a Superior CX

Personalization: Personalization refers to the degree to which data are customized to the needs of a user. It forms a positive determinant of positive experience (Bilgihan et al., 2016). Personalization is one of the key components associated with AI (Zanker et al., 2019). Screen layout, visual elements, animation, and graphical information are some of the areas where a firm can offer personalization and make a difference. Content is curated with the help of customer profiling and psychographics. Personalization helps deliver a richer CX. The 24×7 availability of chatbots plus the addition of a touch of personalization makes the customer feel listened to and happy. Chatbots can identify trigger words from the customer's text and use them aptly to search. The element of personalization is evident from the customized solution offered to customers each time for their different queries. Regarding content delivery, their experience can be enriched by using AI to deliver content based on the stories they interact with the most and by giving them the option to choose how often they want to receive content. It creates a positive brand experience and drives traffic to the website. The common concerns of consumers in the field of AI are the lack of human interaction and an intrusion in privacy. Trust plays an important role in the AI-enabled CX.

Financial Sense: A well-designed chatbot can guide customers through their doubts and convert a prospective customer into a real one by acting as a catalyst during the consumer journey. It certainly helps generate revenue and even cuts down on costs through the automation of the general customer support process.

TABLE 3.1
Industry Examples of Chatbots

Food		Technology		Financial Technology (Fintech)	
Taco Bell	TacoBot	Google	Assistant	Citi Bank	Ask Me
Domino's	Dom	Microsoft	Tay	HSBC	Andrew
TGIF	TGIFridays	Yahoo		HDFC bank	EVA
		Apple	Siri	SBI	SIA
		Tech Mahindra	Entellio	Kotak Mahindra Bank Ltd	Keya
Travel		**Retail**		**Others**	
Emirates Airlines	Emirates vacations	Amazon	Alexa	Microsoft	Xiaoice

Content: Chatbots make natural, lighthearted, and fun conversations with prospective customers – keeping a conversational tone and using emojis. The availability of preset options also makes it easy for customers to find the required information fast. Exercises such as wanting to learn something but wanting an alternative way to do so and providing simple solutions always deliver delight to customers.

Various concerns have been discussed on how customers would react to these chatbots. Brandtzaeg and Følstad (2017) highlighted that chatbots are often used to decrease the customer waiting period and enhance productivity, i.e., in case a chatbot can resolve a query quickly and with precision, customers will automatically gravitate toward such technology. Trivedi (2019) discussed factors like response time, usability, reliability, and availability that impact consumers' experience of using banking chatbots. Indian banking is also witnessing a shift in terms of technology adoption. Many Indian banks such as HDFC, ICICI, and SBI are utilizing chatbots for improved customer service. Table 3.1 provides a list of Chatbots across different industries.

3.5.2 Issues with Chatbots

Chatbots are sequentially scripted with logic and follow a premeditated path of questions exhibiting minimal intelligence (Tuzovic & Paluch, 2018). There are two levels of chatbots, basic and advanced. Basic chatbots can be installed where the requirement is to respond to frequently asked questions (FAQs) (Buhalis & Cheng, 2020). And advanced chatbots are driven by AI. However, AI-driven chatbots function and communicate well as they work through natural language processing (Griol et al., 2013). Consumers risk use in chatbots as they are comparatively novel technology platforms. Therefore, the moderating role of perceived risk between the three quality dimensions and CX is also observed. If a customer gets an enhanced experience, he immediately starts feeling good about the brand and its offerings. During the interaction process, customers come across difficulties such as authenticity issues, cognitive or affective issues, functionality hassles, or integration conflicts which give rise to failed service delivery. Also, chatbots display a lack of understanding if they are loaded with multiple questions (they would respond to all queries in the same manner). Cognitive issues also include incorrect interpretation, i.e., the reply a customer gets is not connected to the query. Another dimension that sometimes appears missing is an element of personalization and human empathy.

3.6 CONCLUSION

CX is getting more fragmented, and the new technologies are rather more complex; technologies that are simpler, cost-effective, and carry human intelligence are needed in the future.

Technology helps in decreasing brand risk. Social media, other communication channels, and enhanced networks should be leveraged by organizations to mitigate risks. Investment in technology needs to be thoughtful, and implementation needs to be thorough.

Firms can well utilize technologies to strengthen their online shopping experience, improve conversion rates, and counter web-rooming behavior. Benefits can be extracted through integration channels such that customers can share the relevant content through social media and other platforms. This shared content itself can act as an influencer and attract other customers. Technologies can be used as a platform for sharing, connecting, supporting, discussing, and interacting with others. Social empowerment boosts positive word-of-mouth marketing and usage intention and helps firms address technology adoption issues. Organizations should promote socially empowering interactions through digital technologies. AR is quite advantageous compared to other media as it can combine audio and video and parallelly combine a live view in real-time with virtual images (Kleef et al., 2010). These interactions allow real-time access to information revolutionizing the world. Humans should control machines to provide dependable and responsible services (Lui & Lamb, 2018).

REFERENCES

Azuma, R. T. (1997). A survey of augmented reality. *Presence: Teleoperators and Virtual Environments, 6*(4), 355–385.

Beck, M. & Crié, D. (2016). I virtually try it ... I want it! virtual fitting room: A tool to increase online and offline exploratory behavior, patronage, and purchase intentions. *Journal of Retailing and Consumer Services, 40.* doi:10.1016/j.jretconser.2016.08.006.

Bilgihan, A., Kandampully, J., & Zhang, T. (2016). Towards a unified customer experience in online shopping environments. *International Journal of Quality And Service Sciences, 8*(1), 102–119. doi:10.1108/ijqss-07-2015-0054.

Brandtzaeg, P. B. & Følstad, A. (2017). Why people use chatbots. In: Kompatsiaris, I. et al. (eds) *Internet Science.* INSCI 2017. Lecture Notes in Computer Science (vol. 10673). Springer, Cham. doi:10.1007/978-3-319-70284-1_30.

Brill, T., Munoz, L., & Miller, R. (2019). Siri, Alexa, and other digital assistants: A study of customer satisfaction with artificial intelligence applications. *Journal of Marketing Management, 35*(15–16), 1401–1436. doi:10.1080/0267257x.2019.1687571.

Buchanan, B. G. (2005). A (very) brief history of artificial intelligence. *AI Magazine, 26*(4), 53.

Buhalis, D., & Cheng, E.S.Y. (2020). Exploring the use of chatbots in hotels: Technology providers' perspective. In: Neidhardt J., & Wörndl W. (eds.) *Information and Communication Technologies in Tourism 2020.* Springer, Cham. doi:10.1007/978-3-030-36737-4_19.

Buhalis, D., & Leung, R. (2018). Smart hospitality—Interconnectivity and interoperability towards an ecosystem. *International Journal of Hospitality Management, 71*, 41–50. doi:10.1016/j.ijhm.2017.11.011

Colon, T. (2018). Eight tips for boosting sales in your digital sales transformation. *Forbes*. https://www.forbes.com/sites/forbestechcouncil/2018/07/13/eight-tips-for-boosting-sales-inyour-digital-sales-transformation/#1a55fe3f52a0.

Davis, F. D. (1989). Perceived usefulness, perceived ease of use, and user acceptance of information technology. *MIS Quarterly, 13*(3), 319–340. doi:10.2307/249008.

Elmorshidy, A. (2013). Applying the technology acceptance and service quality models to live customer support chat for e-commerce websites. *Journal of Applied Business Research (JABR), 29*(2), 589.

Flavián, C., Ibáñez-Sánchez, S., & Orús, C. (2019). The impact of virtual, augmented and mixed reality technologies on the customer experience. *Journal of Business Research, 100*(C), 547–560.

Grewal, D., Levy, M., & Kumar, V. (2009). Customer experience management in retailing: An organizing framework. *Journal of Retailing, 85*(1), 1–14. doi:10.1016/j.jretai.2009.01.001.

Griol, D., Carbó, J., & Molina, J. M. (2013). An automatic dialog simulation technique to develop and evaluate interactive conversational agents. *Applied Artificial Intelligence, 27*(9), 759–780. doi:10.1080/08839514.2013.835230.

Güngör, H. (2020). Creating value with artificial intelligence: A multi-stakeholder perspective. *Journal of Creating Value, 6*(1), 72–85.

Han, D., Tom Dieck, M., & Jung, T. (2019). Augmented Reality Smart Glasses (ARSG) visitor adoption in cultural tourism. *Leisure Studies, 38*(5), 618–633. doi:10.1080/02614367.2019.1604790.

Heller, J., Chylinski, M., de Ruyter, K., Mahr, D., & Keeling, D. I. (2019). Let me imagine that for you: Transforming the retail frontline through augmenting customer mental imagery ability. *Journal of Retailing, 95*(2), 94–114.

Hertzfeld, E. (2019). Japan's Henn na Hotel fires half its robot workforce. *Hotel Management*, 31 January. https://www.hotelmanagement.net/ tech/japan-s-henn-na-hotel-fires-half-its-robot-workforce.

Hilken, T., Keeling, D.I., de Ruyter, K., et al. (2020). Seeing eye to eye: social augmented reality and shared decision making in the marketplace. *Journal of the Academy of Marketing Science 48*, 143–164. doi:10.1007/s11747-019-00688-0

Hill, J., Randolph Ford, W., & Farreras, I. (2015). Real conversations with artificial intelligence: A comparison between human–human online conversations and human–chatbot conversations. *Computers in Human Behavior, 49*, 245–250.

Holder, C., Khurana, V., Harrison, F., & Jacobs, L. (2016). Robotics and law: Key legal and regulatory implications of the robotics age (Part I of II). *Computer Law & Security Review, 32*(3), 383–402.

Holloway, B., Wang, S., & Parish, J. (2005). The role of cumulative online purchasing experience in service recovery management. *Journal of Interactive Marketing, 19*(3), 54–66. doi:10.1002/dir.20043.

Huang, Y., Li, H., & Fong, R. (2015). Using Augmented Reality in early art education: A case study in Hong Kong kindergarten. *Early Child Development and Care, 186*(6), 1–16.

Jain, R., Aagja, J., & Bagdare, S. (2017). Customer experience – A review and research agenda. *Journal of Service Theory and Practice, 27*(3), 642–662.

Javornik, A. (2016). Augmented reality: Research agenda for studying the impact of its media characteristics on consumer behavior. *Journal of Retailing and Consumer Services, 30*. doi:10.1016/j.jretconser.2016.02.004.

Jones, L., Golan, D., Hanna, S., & Ramachandran, M. (2018). Artificial intelligence, machine learning and the evolution of healthcare. *Bone & Joint Research, 7*(3), 223–225. doi:10.1302/2046-3758.73.bjr-2017-0147.

Kalantari, M., & Rauschnabel, P. (2018). Exploring the early adopters of augmented reality smart glasses: The Case of Microsoft HoloLens. Progress in IS. In: Jung, T., & Claudia tom Dieck, M. (eds.), *Augmented Reality and Virtual Reality*, 229–245, Springer.

Kaushal, V. & Yadav, R. (2020). Understanding customer experience of culinary tourism through food tours of Delhi. *International Journal of Tourism Cities*. doi:10.1108/IJTC-08-2019-0135.

Kim, J., Jin, B., & Swinney, J. L. (2009). The role of etail quality, e-satisfaction and e-trust in online loyalty development process. *Journal of Retailing and Consumer Services, 16*(4), 239–247. doi:10.1016/j.jretconser.2008.11

Kleef, N., Noltes, J., & Spoel, S. (2010). *Success factors for Augmented Reality Business Models*. University of Twente, Enschede.

Kumar, A. & Paul, J. (2018). Mass prestige value and competition between American versus Asian laptop brands in an emerging market-theory and evidence. *International Business Review, 27*(5), 969–981.

Larivière, B., Bowen, D., Andreassen, T. W., Kunz, W., Sirianni, N. J., Voss, C., Wünderlich, N. V., & De Keyser, A. (2017). "Service encounter 2.0": An investigation into the roles of technology, employees, and customers. *Journal of Business Research, 79*, 238–246.

Leung, R. & Loo, P. (2020). Co-creating interactive dining experiences via interconnected and interoperable smart technology. *Asian Journal of Technology Innovation*, 1–23.

Lui, A. & Lamb, G. (2018). Artificial intelligence and augmented intelligence collaboration: regaining trust and confidence in the financial sector. *Information & Communications Technology Law, 27*(3), 267–283. doi: 10.1080/13600834.2018.1488659.

Makadia, M. (2018). How cutting-edge hotels use artificial intelligence for a great guest experience. *PhocusWire*. Retrieved from https://www.phocuswire.com/How-cutting-edge-hotelsuse-artificial-intelligence-for-a-great-guest-experience.

Marasco, A., Buonincontri, P., van Niekerk, M., Orlowski, M., & Okumus, F. (2018). Exploring the role of next-generation virtual technologies in destination marketing. *Journal of Destination Marketing & Management, 9*, 138–148.

McLean, G. & Wilson, A. (2019). Shopping in the digital world: Examining customer engagement through augmented reality mobile applications. *Computers in Human Behavior, 101*, 210–224.

Miner, A. S., Laranjo, L., & Baki. (2020). Chatbots in the fight against the COVID-19 pandemic. *npj Digital Medicine, 3*(1), 1–4.

Moriuchi, E., Landers, V. M., Colton, D., & Hair, N. (2020). Engagement with chatbots versus augmented reality interactive technology in e-commerce. *Journal of Strategic Marketing, 29*(11), 1–15. doi:10.1080/0965254X.2020.1740766.

Nanyang Technological University Singapore. (2019). Nadine social robot. https://imi.ntu.edu.sg/IMIResearch/ResearchAreas/Pages/NadineSocialRobot.aspx.

Oram, R. (2019). Meeting Edward: Chatbots and the changing the face of the hotel guest experience. *Oracle Hospitality Check-In*. https://blogs.oracle.com/hospitality/chatbots-and-the-changing-theface-of-the-hotel-guest-experience.

Osawa, H., Ema, A., Hattori, H., Akiya, N., Kanzaki, N., Kubo, A., Koyama, T., & Ichise, R. (2017). What is real risk and benefit on work with robots? *Proceedings of the Companion of the 2017 ACM/IEEE International Conference on Human-Robot Interaction*, Vienna, Austria.

Ostrom, A. L., Parasuraman, A., Bowen, D. E., Patricio, L., & Voss, C. A. (2015). Service research priorities in a rapidly changing context. *Journal of Service Research, 18*(2), 127–159.

Poushneh, A. (2018). Augmented reality in retail: A trade-off between user's control of access to personal information and augmentation quality. *Journal of Retailing and Consumer Services, 41*, 169–176.

Rauschnabel, P. A., Felix, R. & Hinsch, C. (2019). Augmented reality marketing: How mobile AR-apps can improve brands through inspiration. *Journal of Retailing and Consumer Services*, 49(C), 43–53.

Rauschnabel, P. A., Rossmann, A., & tom Dieck, M. (2017). An adoption framework for mobile augmented reality games: The case of Pokémon Go. *Computers in Human Behavior*, 76, 276–286. doi:10.1016/j.chb.2017.07.030.

Rese, A., Baier, D., Geyer-Schulz, A., & Schreiber, S. (2017). How augmented reality apps are accepted by consumers: A comparative analysis using scales and opinions. *Technological Forecasting And Social Change*, 124, 306–319.

Riener, A., Gabbard, J., & Trivedi, M. (2019). Special issue of presence: Virtual and augmented reality virtual and augmented reality for autonomous driving and intelligent vehicles: Guest editors' introduction. *PRESENCE: Virtual and Augmented Reality*, 27(1), i–iv.

Roe, D. (2017). Forrester wave finds 3 trends reshaping sales force automation. Accessed January 1, 2020. https://www.cmswire.com/digital-marketing/forrester-wave finds-3-trends-reshaping-sales-force-auto.

Rose, S., Clark, M., Samouel, P., & Hair, N. (2012). Online customer experience in e-Retailing: An empirical model of antecedents and outcomes. *Journal of Retailing*, 88(2), 308–322.

Schlicht, M. (2016). The complete beginner's guide to chatbots. *Chatbots Magazine*. Retrieved from https://chatbotsmagazine.com/the-complete-beginner-sguide-to-chatbots8280b7b906ca#.5vs9cl5ut.

Scholz, J. & Smith, A. (2016). Augmented reality: Designing immersive experiences that maximize consumer engagement. *Business Horizons*, 59(2), 149–161.

Solomon, M. (2016). Technology invades hospitality industry: Hilton robot, Domino delivery droid, Ritz-Carlton mystique. *Forbes*. Retrieved from https://www.forbes.com/sites/micahsolomon/2016/03/18/high-tech-hospitality-hilton-robot concierge-dominos-delivery-droid-ritz-carlton-mystique.

Trivedi, J. (2019). Examining the customer experience of using banking chatbots and its impact on brand love: The moderating role of perceived risk. *Journal of Internet Commerce*, 18(1), 91–111.

Tuzovic, S. & Paluch, S. (2018). Conversational commerce – A new era for service business development? In: Bruhn, M. & Hadwich, K. (eds) *Service Business Development* (pp. 81–100). Springer Gabler, Wiesbaden.

Verhoef, P., Lemon, K., Parasuraman, A., Roggeveen, A., Tsiros, M., & Schlesinger, L. (2009). Customer experience creation: Determinants, dynamics and management strategies. *Journal of Retailing*, 85(1), 31–41. doi:10.1016/j.jretai.2008.11.001.

Wolfe, C. (2017). Global study from Cohn & Wolfe defines authenticity in the eyes of consumers and reveals the 100 most authentic brands. Retrieved January 5, 2021, from https://www.prnewswire.com/news-releases/global-study-from-cohn--wolfe-defines-authenticity-in-the-eyes-of-consumers-and-reveals-the-100-most-authentic-brands-300253451.html

Yim, M. Y. C., Chu, S. C., & Sauer, P. L. (2017). Is augmented reality technology an effective tool for eCommerce? An interactivity and vividness perspective. *Journal of Interactive Marketing*, 39, 89–103.

Zanker, M., Rook, L., & Jannach, D. (2019). Measuring the impact of online personalization: Past, present and future. *International Journal of Human-Computer Studies*, 131, 160–168. doi:10.1016/j.ijhcs.2019.06.006.

Zhou, F., Duh, H. B. L., & Billinghurst, M. (2008). Trends in augmented reality tracking, interaction and display: A review of ten years of ISMAR, *2008 7th IEEE/ACM International Symposium on Mixed and Augmented Reality*, Cambridge, UK, pp. 193–202. doi:10.1109/ISMAR.2008.4637362.

4 Use of Artificial Intelligence-Enabled Features in the Retail Sector
A Perceptual Study of Customers

Ashutosh Mohan, Upnishad Mishra, and Ishi Mohan
Banaras Hindu University (BHU)

CONTENTS

4.1 Introduction	62
4.2 Artificial Intelligence	63
4.2.1 Different Stages of AI	64
4.2.2 Different Components of AI	66
4.3 Machine Learning	66
4.3.1 The Working Process of Machine Learning Model	67
4.3.2 Machine Learning Methods	69
4.4 Deep Learning	69
4.5 Emerging Artificial Intelligence Technologies	72
4.5.1 Explainable AI	72
4.5.2 Leading AI Companies	72
4.6 AI and Retail	77
4.6.1 Digital Giants and AI	79
4.7 AI Tools	81
4.7.1 Chatbots	81
4.7.2 Personalized Marketing	83
4.7.3 Marketing and Content Automation	84
4.7.4 Voice Assistance	85
4.7.5 Augmented Reality	86
4.7.6 Visual Search	87

DOI: 10.1201/9781003140474-4

4.8   Methodology 88
    4.8.1   Data Analysis and Results 88
4.9   Conclusion and Future Trends 90
References 91

4.1   INTRODUCTION

The fourth industrial revolution powered by IoT, big data, machine learning, and artificial intelligence (AI) is changing the market dynamics swiftly. These subsequent market changes have a profound impact on both the marketers and the customers. Technology such as artificial intelligence (AI) has an enormous potential for vastly revamping the current marketing practices. AI offers entirely new ways of creating and distributing value to marketers, such as personalized marketing, predictive analytics, content automation, and in-depth insights, which allow for a more comprehensive view of customer behavior. Similarly, these AI platforms and assistants offer highly personalized and scalable experiences for the customers and likewise change the way customers interact with a marketplace. It also tends to reconfigure the market dynamics and reshape the companies that sell into it.

"Artificial Intelligence means any technology that seeks to mimic human intelligence, which covers a huge range of capabilities such as voice and image recognition, machine learning techniques, and semantic search" (Robert Allen, 2017). Almost every sector has seen the integration of AI in their various business processes. "The recent explosion of advances in the use of AI is changing the way many problems are being solved to the extent that one can say the world is experiencing a renaissance" (Tan and Lim, 2018). The worldwide spending on AI was expected to reach about $19.1 billion in 2018 according to Business Wire, and according to the forecast of worldwide semiannual cognitive artificial intelligence system spending guide of the international data corporation (IDC, 2016), the compound annual growth rate (CAGR) of spending in AI, it will grow 46.2% throughout 2016–2021, and the spending is expected to reach $52.2 billion in 2021. According to the same report, retail will lead in AI spending for product recommendations, personalized marketing, and automated customer support services, followed by banking, manufacturing, and healthcare sectors. According to the 2020 sixth state of marketing report by salesforce, based on a survey of 7,500 marketers worldwide, 84% of marketers reported using AI. It is difficult to ignore the AI in the present scenario because of the advantage it provides to the businesses. Companies such as Google, Flipkart, Amazon, etc., are using AI-enabled technologies to target their customers more effectively and provide better services. AI is being used for enhanced marketing operation, reliable CRM solutions, and effective delivery of services to customers. AI-based technologies such as predictive solutions, based on the propensity model, help predict the customers' requirements or prospective customers. It is being used to provide personalized solutions to the customers, providing them with more relevant options and ultimately increased sales. AI features enabled with machine learning and natural language processing prove to provide enhanced experiences and augmented features.

A majority of research has been done on the technical aspects of AI (Wierenga et al. 2010); customer perception toward the online retail sector's AI features becomes a crucial topic to do research. Potentially, AI can make an essential contribution to marketing decision making. Various studies have been done on privacy and trust issues (Hoy, 2018), experience (Brandtzaeg & Følstad et al., 2017), and motivation (Watson et al., 2018; Chopra, 2019), derived from the use of different AI features. A gap is identified when one tries to find out the study on consumer satisfaction (Riikkinen et al., 2018) related to these AI features. The present study explores the online retail sector trends, particularly how the fourth industrial revolution, mainly AI, is changing the marketing dynamics. The study uses data collected from online buyers to analyze which AI-based feature is significant for the consumer and how much they prefer using them. By analyzing its current applications and customer perception, one can acquire a high-level understanding of AI's long-term implications in the field of marketing and its areas for improvement.

4.2 ARTIFICIAL INTELLIGENCE

In 1950, while working at the University of Manchester, Allen Turing published a paper titled "computing machinery and intelligence," which gave the famous Turing test. They tried to address the two central questions in this paper: first was "can machines think?" and the second question was "are there imaginable digital computers which would do well in the imitation game?" A year later, while doing his Ph.D. at Princeton, Marvin Minsky built SNARC, the first artificial neural network computer. It was the first incidence when the concept of a machine, which can think like a human, was proposed. After few years, John McCarthy, who is also known to be the father of AI, first, coined the term artificial intelligence in the famous Dartmouth Conference in 1956. McCarthy et al. (1995) described artificial intelligence as "making a machine behave in ways that would be called intelligent if a human were so behaving." In contrast, the cognitive scientist Minsky (1974) defined AI as "the science of making machines do things that would require intelligence if done by men." Stone et al. (2016) quoted Nils J. Nilsson's definition: "Artificial intelligence is that activity devoted to making machines intelligent, and intelligence is that quality that enables an entity to function appropriately and with foresight in its environment." According to Castro and New (2016), "most AI have at least one of seven functions: monitoring; discovering; predicting; interpreting; interacting with the physical environment; interacting with humans; and interacting with machines." Rao (2017) put forward Russell and Norvig's (1995) definition of AI as "the designing and building of intelligent agents that receive percept from the environment and take actions that affect that environment." Different researchers defined the term AI differently, and there seems no clear consensus among the AI developers on one definition of AI (Simon, 2019).

During the initial years, AI developed significantly, but from 1980 to 1990, AI development became stagnant; it was termed as AI winters. That was when nothing substantial happened in AI due to negligence and drop in investment by the government and the private players into this field and the researcher's pessimistic approach toward AI. During the late 20th century, the AI technology concerned with learning algorithms' design could process only a small data set and hence had limited capabilities and application. After a long spell of AI winters, the field of AI saw a dramatic change (Figure 4.1).

FIGURE 4.1 The AI timeline.

With the increase in the popularity and wide use of computers and the internet, AI's scope increased. Stakeholders felt the importance of AI, and the possible benefits of its use, investment, and research were understood. The periods of 1956–1980 and 2011–2017 were described as the two "hype cycles" in AI. Aggarwal (2018) described the hype cycle as a time period when the developers, researchers, and investors become optimistic about the AI and enormous growth occurs. Three primary reasons for AI's rejuvenation were believed to be "the availability of big data, supercomputing power to process it over large neural networks and modern algorithms" (Forbes, 2019) (Table 4.1).

Breakthrough came in AI with the enhancement in the capacity of extensive data set processing by machine learning. It gave the AI agents the ability to correctly interpret large data sets, learn from such data, and use those learning to achieve specific goals (Kaplan and Haenlein, 2018). Further machine learning and deep learning have facilitated object recognition in images, video labeling, natural language processing, leading to audio and speech-based solutions for customers, and broader acceptability of the technology (Varian, 2018). In 2011 IBM launched its AI-based platform WATSON, and in the same year, Apple launched SIRI, a voice assistant platform for its users. In 2014, Amazon launched a similar AI-enabled platform named Alexa, including voice assistant, personalized product recommendation, etc. In 2016, the company Deep Mind, an AI start-up, developed an AI-based computer program, AlphaGo Art, which achieved global coverage after defeating the existing Chinese board game "Go" world champion Se-dol Lee, in Seoul.

4.2.1 Different Stages of AI

Any machine or tool can be AI-enabled if it possesses any one or more than one feature similar to human intelligence, such as vision, reasoning, planning, learning from experience, and modifying itself according to its surroundings. According to Kaplan and Haenlein (2018), AI can be divided into three different stages based on

TABLE 4.1
Major Breakthroughs in the Field of AI

Year	Breakthroughs in AI	Data sets (First Available)	Algorithms (First Proposed)
1994	Human-level spontaneous speech recognition	Spoken wall street journal articles and other texts (1991)	Hidden Markov model (1984)
1997	IBM Deep Blue defeated Garry Kasparov	700,000 Grandmaster chess games, "The Extended Book" (1991)	Negascout planning algorithm (1993)
2005	Google's Arabic and Chinese to English translation	1.8 trillion tokens from Google web and new pages (collected in 2005)	Statistical machine translation algorithm (1988)
2011	IBM Watson became the world jeopardy champion	8.6 million documents from Wikipedia, Wikitionary, Wikiquote, and project Guttenberg (updated in 2010)	Mixture-of-experts algorithm (1989)
2014	Google's GoogleNet object classification at near-human performance	ImageNet corpus of 1.5 million labeled images and 1,000 object categories (2010)	Convolution neural network algorithm (1991)
2015	Google's Deepmind achieved human parity in playing 29 Atari games by learning general control from video	Acrade learning environment data set of over 50 Atari games (2013)	q-learning algorithm (1992)

Source: NITI Aayog (2018).

the level of human intelligence the machine or tool possesses, and the same was explained through a model given in Figure 4.2, which uses the example of Apple's Siri to explain the same.

Three different stages of AI are as follows:
  a. **Artificial Narrow Intelligence (ANI)**: ANI may be categorized as the first generation of AI designed to carry out specific and predetermined tasks. It is weak and possesses below human-level intelligence, such as Google assistance or Facebook's face recognition technique for tagging users. The scope of ANI is minimal.
  b. **Artificial General Intelligence (AGI)**: It will be the second generation of AI that will perform tasks like humans. They will be able to plan, reason, and autonomously solve the problems and make decisions they were not even designed for.
  c. **Artificial Super Intelligence (ASI)**: According to Kaplan and Haenlein (2018), "we might see the third generation, artificial super intelligence (ASI), which is truly self-aware and conscious systems that, in a certain way, will make humans redundant." These systems will have the wisdom and ability of scientific creation and may even perform those beyond human abilities.

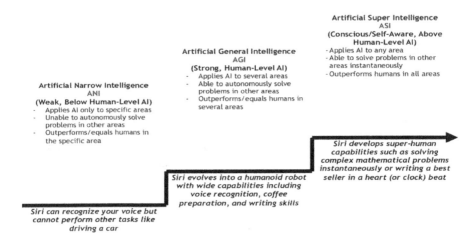

FIGURE 4.2 Different stages of AI. (Kaplan and Haenlein, 2018.)

Though the AI has made a significant development, experts still believe that it is far from achieving a human brain's capability. According to Wharton (2017), "the current AI level is like what he calls the "old brain," similar to the cognitive ability of rats." As regards "strong AI," often termed as AGI, there is some consensus that we may be "a long way from artificial general intelligence," at least by a decade (Brynjolfsson et al., 2017; NSTC report, White House, 2016), and ASI is far from achievable. The Dartmouth College Artificial Intelligence Conference: The Next 50 Years (AI@50) was held in July 2006, at the famous Dartmouth College, where the term AI was first coined in 1956, to assess how far AI has progressed and the road ahead. Many predictions of AI were made at the conference. According to McCarthy, the father of AI, we may achieve a human-level AI but not assured by 2056. In contrast, Selfridge claimed that the developers might incorporate feelings in the AI system by the next 50 years but will not achieve a human-level AI.

4.2.2 DIFFERENT COMPONENTS OF AI

AI is a combination of several components that together make AI possible. "AI is not a product, like packaged software, mobile apps or operating systems. It combines software and hardware, but there is no specific AI software, just AI solutions which build upon a variety of software" (Simon, 2019). AI systems are not universal; they are designed to accomplish a particular task. The diagram given below describes the two different subsets or components of AI. Three different terms discussed in the following diagram are AI, machine learning, and deep learning, which are inter-related and make the AI possible (Figure 4.3).

4.3 MACHINE LEARNING

Artur Samuel (1959) was the first to introduce the term machine learning. It means "the ability to learn without being explicitly programmed." It is a subset of AI that uses statistical method, which helps extract patterns from data sets to create

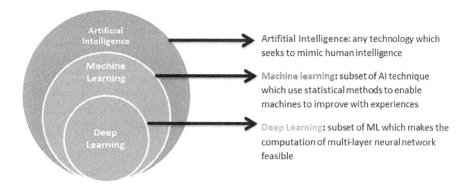

FIGURE 4.3 Different components of artificial intelligence.

algorithms, enabling machines to improve with experiences. Instead of hand-coded instructions for a specific task, a machine is trained on large amounts of data, which gives the machine the ability to learn from such data and perform specific tasks. The machine learning model is basically based on deep learning and consists of multiple artificial neural networks with more than one hidden layer in the model.

4.3.1 THE WORKING PROCESS OF MACHINE LEARNING MODEL

The machine learning application (or model) is developed in four steps. The data scientist developing the model works with the partners for whom the model is being developed to get the desired output from the model. A machine learning model can be developed by following the four step procedure.

Step 1: Selecting the training data

The machine learning model is first run on the training data. Training data represent how the model has to be run later to solve the required problem. The training data can be of two types: labeled and unlabeled. The labeled data are those where data are tagged; to mark the features, the model needs to recognize and classify based on the same. The other form of data is unlabeled or untagged data, where the model has to identify different attributes and assign classification on its own. The data can be further divided into the training subset and the evaluation subset used to train, test, and refine the model. In both cases, data need to be adequately prepared to avoid any biases or imbalances that could impact the machine learning model.

Step 2: Developing an algorithm to run the training data

An algorithm is a statistical formula or a process consisting of a sequence of specified actions to solve a problem. In a machine learning model, the algorithm depends on various factors such as the nature of the problem, labeled or unlabeled data, and the training data set and its authenticity. Common types of machine learning algorithms for use with labeled data include the following:

a. **Regression Algorithm**: To understand the relationship in data regression algorithms is used. Linear regression is used when the developer needs to predict any dependent variable's value based on an independent variable's value. For example, to predict the company's annual sales (dependent variable) based on its advertisement expense (independent variable), a linear regression algorithm can be used. Logistic variable is used in any model when the dependent variable is binary such as pass or fail, X or Y, etc. whereas, a different kind of regression algorithm known as support vector is used when the dependent variable is not easy to classify.
b. **Decision Trees**: In the decision tree algorithm model, classified data are used by the system to make recommendation based on certain predetermined rules. For example, a model based on a decision tree algorithm, which predicts an individual football team's winning, uses the team's data, such as past performances, players, venue, etc. and applies the rules to these factors to suggest an action or decision.
c. **Instance-Based Algorithms**: In a machine learning model, an instance-based algorithm compares new problem instances with the training instances and makes forecast on the basis of the same. The instance-based algorithm model uses categorization to evaluate the probability of a data to be a member of a particular group based on its close relationship with the group members. K-Nearest Neighbor or k-nn are the two examples of an instance-based algorithm model.

Algorithms to be used with unlabeled data include the following:
a. **Clustering Algorithms**: These algorithms strive to identify natural grouping in any data. The model does not have any prior knowledge or the characteristics of the group. It identifies the similar kind of records from the data and labels them accordingly. Kohonen and K-means are a few examples of clustering algorithms.
b. **Association Algorithms**: These algorithms are used for discovering patterns and relations between the variables in databases. A simple "if-then" relationship is used to find a relationship called association, similar to data mining rules.
c. **Neural Networks**: A neural network is a series of algorithms, which consist of different layered networks. The network has an input layer used to feed the data; at least one hidden layer, where the data ingested in the network are processed, all the necessary calculations are performed, and the required result is generated; and an output layer. A neural network with multiple hidden layers between the input and the output is called a deep neural network.

Step 3: Algorithm training for finalizing the model

The third step is concerned with the iterative processing of the training algorithm by running the variable through it. Then the output produced by the algorithm is compared with the desired output. If any variation in the result is found, variable weights are adjusted, and possible biases are checked within the system to provide

Use of AI-Enabled Features in Retail

a more accurate outcome. The process is repeated again and again till the time the model produces the desired result frequently.

Step 4: Using the model and making the required improvements

The last step is concerned with the use of the machine learning model by feeding new data into it. The model is used continuously, and the required improvements are made from time to time to improve its efficiency and accuracy. The source of the new data largely depends on the problem being solved. For example, a machine learning model designed to predict the weather will use data from various satellites, websites, and other different sources, whereas a system developed for identifying spam mails will ingest emails.

4.3.2 Machine Learning Methods

Supervised, unsupervised, and reinforcement learning are the three different machine learning methods or styles. Supervised machine learning uses a labeled or tagged input data set. For example, a supervised model developed to recognize a flower such as a rose in an image is trained on a data set of various tagged flower images. This kind of model generally requires less training, and training is easier as labeled data are used. One of the drawbacks of this method is that an excellent labeled data without any biases is expensive and not easy to prepare.

The second method is unsupervised machine learning. This method is used to classify the data and find any possible relationship or pattern in them. The method uses unlabeled or untagged data to extract the information without any human intervention. For example, an unsupervised learning algorithm can sort spam messages from billions and billions of emails worldwide. The third form of machine learning method is reinforcement learning, where the model learns using the trial and error method. A sequence of successful results is reinforced to develop the best possible recommendation for a given problem. One of the good examples of the reinforcement learning model is IBM Watson. Different methods of machine learning and their possible uses have been described in Table 4.2.

The most recent advancement in machine learning technology was its ability to process large data sets, making its wider accessibility possible. Online retailers use machine learning technology to make predictions and recommendations to the users by experience and detecting patterns in the data. The algorithms further adapt to new data and experiences to improve the online platform's efficiency and effectiveness over time.

4.4 DEEP LEARNING

It is a subset or a specific class of machine learning algorithms. It consists of a multilayer of artificial neural network (ANN) interconnecting many neurons inspired by the human brain's structure and functioning. Similar to the human neuron system, ANNs have discrete layers and connections to other artificial neurons. Each layer is designed to pick out a specific feature to learn. It is this multilayer of neurons that gives deep learning its name. At its most basic, a neural network has an input layer used to feed the data and at least one hidden layer, where the data ingested in

TABLE 4.2
Machine Learning Methods

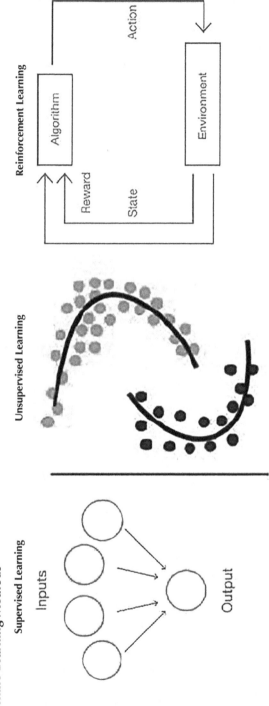

	Supervised Learning	Unsupervised Learning	Reinforcement Learning
What is it?	Supervised learning methods are based on the algorithms which use feedback and the input data provided by the humans. The model is then programmed to learn the relationship between the input data and the output (e.g., how the input "demand" and the "supply" predict the prices)	Unsupervised machine learning method is able to identify the different pattern or relationship from the input data without being explicitly programmed (e.g., customer segmentation by the model by using the demographic data of the customers)	Reinforcement learning is a machine learning model which learns to execute an action by trying to maximize the notion of cumulative reward (example of reinforcement learning is maximum points the system attains for increasing returns on financial investments)

(*Continued*)

TABLE 4.2 (Continued)
Machine Learning Methods

	Supervised Learning	Unsupervised Learning	Reinforcement Learning
When to use it	The particular method is used when you can identify the different factors and able to classify the input data. You are also familiar with the behavior or output you want to predict using the system, but want the model to predict it for you on the new data set	This method is used when you are dealing with unlabeled data where the model is responsible for identifying the pattern and classify the data	The reinforcement learning method is used when you do not have an adequate amount of data to train the algorithm. To learn about the environment, the model needs to interact with it and make the predictions or suggestions based on the learning from these interactions
How it works	1. The input data is fed externally in the system (e.g., in the case of predicting prices, labels the input data as "demand", "supply", etc. and defines the output variable, e.g., "prices") 2. The system is trained on the available data to find the relationship between the input variable and the output 3. The system is believed to be trained enough when it achieves a high level of accuracy and is ready to be applied on new data	1. The unlabeled data are provided to the system (e.g., a new data set regarding the email received by an individual) 2. It infers a structure from the data 3. The algorithm identifies the structure and patterns from the available data (e.g., from the millions of mail the algorithm identifies the group of mail to be categorized as spam, social or important)	1. The algorithm interacts with the environment (e.g., makes a financial investment) 2. The system receives reward if the particular trade increases the overall profit (e.g., the highest total returns on all the investments) 3. The algorithm optimizes itself for a particular decision making by correcting itself over time

Source: NITI Aayog (2018).

the network are processed, all the necessary calculations are performed, and the required result is generated an output layer.

Though deep learning technology was developed in the mid-1990s, its practical use was impossible. The use of deep learning technology gained significant importance in AI after developing graphic processing units (GPU) and with the rise of computing power. The necessity of having a solution for processing a vast amount of data generated through various sources made the application of deep learning famous in AI. The neural network can process a large amount of input data through multiple layers to provide the desired result. Deep learning uses many labeled data to process and detect objects in images, speech recognition, language translation, and decisions. Online retailers are using this technology for providing voice-based solutions such as Alexa or Google assistant and image searching, for example, Google lens. Different neural networks and their possible uses have been described in Table 4.3.

4.5 EMERGING ARTIFICIAL INTELLIGENCE TECHNOLOGIES

The field of AI is developing very fast. As defined by Agrawal (2018), it is the second hype cycle we are going through. Developers and investors are very optimistic and putting many resources into developing this technology. Uses of various AI technologies such as computer vision, audio, and natural language processing have grown significantly in the past few years. Some of these emerging technologies in the field of AI have been described in Figure 4.4. (Table 4.4).

4.5.1 EXPLAINABLE AI

One of the evolving research areas in AI that received much attention from the research community is Explainable Artificial Intelligence (XAI). DARPA is working on one such project, which is based on XAI. The XAI by DARPA has shown significant progress in explaining how and why machine learning algorithms work in a certain way. The XAI project aims to produce more explainable machine learning models which "enable human users to understand, appropriately trust, and effectively manage the emerging generation of artificially intelligent partners" (NITI Aayog, 2018). It is believed that the explainable artificial intelligence will bring the third-wave AI systems, where machines will understand the context and the environment in which they are operating and, over time, will be capable enough to characterize real-world phenomena.

4.5.2 LEADING AI COMPANIES

Google, Amazon, and Facebook are some of the leading companies in AI development and usage. Corporate giants like Salesforce, Yahoo, IBM, and Apple are competing to acquire the small AI companies. AI-related financing, marketing, and advertising activities have reached unprecedented levels worldwide. According to McKinsey's estimates, tech giants' global expenditure was between $26 and $39 billion in 2016 on R&D, and AI companies' acquisition and start-ups invested between $6 and $9 billion (McKinsey Global Institute (MGI), 2017a, b) (Table 4.5).

A range of AI-based start-ups is generating funds for various applications. Figure 4.5 shows the external investment in AI-focused companies in 2016. More

TABLE 4.3
Deep Learning and Its Uses

	Convolutional Neural Network	Recurrent Neural Network
	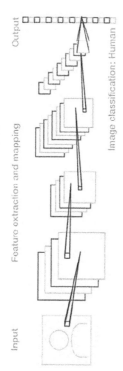	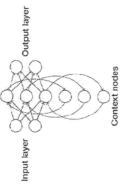
What is it?	A convolutional neural network is deep learning multilayer network architecture to extract complex features of the data at each layer to determine the output	A recurrent neural network is a multilayer deep learning network that can store information in context nodes, which enables it to learn from the data sequences and provide an output in the form of sequences or number
When to use it	This kind of neural network is used when you have unstructured data such as image and you need to acquire information from the same	The particular method is used when you are dealing with the sequence or time searies data such as audio recording
How it works	Processing an image 1. The convolutional neural network (CNN) receives an image, for example, of the letter "A", which it processes as a collection of pixels 2. In the hidden, inner layer of the model, it identifies unique features, for example, the individual line that make up "A" 3. The CNN can now classify a different image as the letter "A" if it finds in it the unique features previously identified as making up the letter	Predicting the next word in the sentence "Are you free…?" 1. A recurrent neural network (RNN) neuron receives a command that indicates the start of the sentence 2. The neuron receives the word "Are" and then outputs the vector of numbers that feeds back into the neurons to help it "remember" that it received "Are" (and that it received first). The same process occurs when it receives "you" and "free", with the state of neuron updating upon receiving each word 3. After receiving "free", the neuron assigns a probability to every word in the English vocabulary that could complete the sentence. If trained well, the RNN will assign the word "tomorrow" one of the highest probabilities and will choose it to complete the sentence

Source: NITI Aayog (2018).

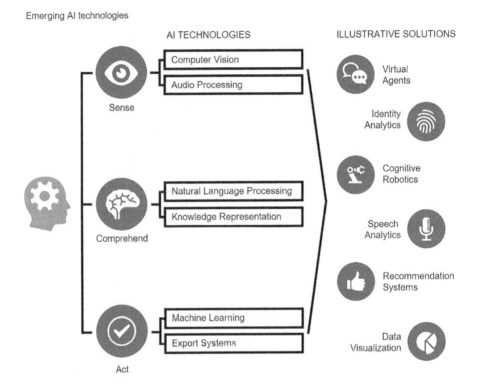

FIGURE 4.4 Emerging AI technologies. (NITI Aayog, 2018.)

TABLE 4.4
Emerging AI Technologies

Computer vision	Computer vision can be defined as "the ability for computers to see imagery through mathematical representations of three-dimensional shape and appearance" (Szeliski, 2011). It gives the system the ability to comprehend the meaning and context of any image similar to humans. It is one of the most prominent forms of machine perception. The current research in the area is focused on automatic image and video captioning
Natural language processing	NLP is the subfield of AI, which is capable of understanding human language. NLP is based on machine learning technology, which gives the voice assistance device the ability to convert human voice into data. "It is this category of AI that will enable computers to understand the hierarchical structure of language and how components of a sentence relate to each other" (Jurafsky and Martin, 2014). The NLP development has opened a wide range of possibilities such as speech to text, grammar correction, modern chatbots, and much more. The recent improvement in NLP is the development in machine learning technology, increased computing power, and linguistic data availability. The AI-based voice assistance such as Alexa and Siri are based on this NLP technology

(Continued)

TABLE 4.4 (*Continued*)
Emerging AI Technologies

Speech recognition	Speech recognition is the technology that is used to decode and convert the human voice into text. Speech recognition is often confused with voice recognition, which is concerned with identifying the individual's voice. An example of speech recognition is Google text to speech, conversion of voice into text messages, and voice-driven phone answering menus
Object identification and image recognition	"Object detection" and "image recognition" are closely related but are two different technologies. Object detection is concerned with identifying a specific object in any image. For example, if we want to know the exact location of a car in any picture, the system will create an output image with a car enclosed in a box with its location. At the same time, image recognition can identify and label the subject matter in any image. Image recognition will create an output with enclosed boxes around the car and label the other objects in the image, such as trees, bikes, and signboards. The technology is used for self-driving cars, medical image analysis, and fingerprint ID systems
Reinforcement learning	Reinforcement learning is a framework where the machine learning model learns using the trial and error method. The best possible solution of a problem is achieved by reinforcing the sequence of successful results. It is a framework that shifted the developer community's attention from pattern detection to sequential decision making. One of the good examples of the reinforcement learning model is IBM Watson
Collaborative systems	The idea behind creating a collaborative system is to develop algorithms and models which can make an autonomous system work in coordination with other system and humans
Algorithmic game theory	It combines game theory and computer science to draw AI technology's attention toward the economic and social computing dimension
Internet of Things (IoT)	IoT is concerned with devices such as cameras, home appliances, or different gadgets that are connected and share their information using sensors
Neuromorphic computing	Carver Mead developed neuromorphic computing in the late 1980s. The technology seeks to mimic the human brain's neural network function to improve the efficiency of the hardware and computer system. The research in this area is focused on achieving the capability of the human brain

Source: Stone et al. (2016), Varian (2018), and Simon (2019).

than 56% of the company's investment in AI was in "machine learning" technology. Twenty-eight percent of the investment was made in "computer vision", followed by 7% in "natural language processing", 6% in "autonomous vehicles", and the rest in "smart robotics" and virtual investment. Investment in the field of AI has more than doubled between 2016 and 2017 to over $15 billion (GSMA, 2018). Different companies are using AI algorithms for different purposes. Details of some of the leading artificial intelligence-based companies are given below.

The top AI companies (Datamation, 2017) are as follows:

i. **AIBrain**: AIBrain is a company based in California, USA. It is a company that builds AI-based products for smartphones and robotic applications. The company strives to provide an AI-based solution with the human skill

TABLE 4.5
External Investments in AI-Focused Companies by Technology Category in 2016

Technology Category	Amount ($ billion, estimates*)
Machine learning multiuse and nonspecific applications	5–7
Computer vision	2.5–3.5
Natural language	0.6–0.9
Autonomous vehicles	0.3–0.5
Smart robotics	0.3–0.5
Virtual agents	0.1–0.2

Source: McKinsey Global Institute (MGI, 2017a, b).

Note: Estimates consist of annual VC investment in AI-focused companies, PE investment in AI-related companies, and M&A by corporations. It includes only disclosed data available in databases and assumes that all registered deals were completed in the year of transaction.

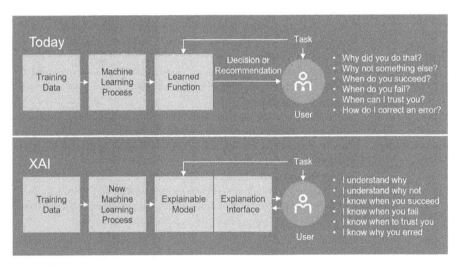

FIGURE 4.5 Explainable artificial intelligence. (DARPA as given in NITI Aayog Discussion paper on AI for all, 2018.)

set of learning and problem-solving. The company provides AI agent, iRSP, and AICoRE, an intelligent robot software platform.

ii. **Amazon***
iii. **Anki**: Anki is a technology company dedicated to bringing robotics into the consumer's everyday life. It is doing so with its two products, i.e., it is "Cozmo" and "Anki Overdrive products". Cozmo is Anki's flagship robot, capable of providing emotional responses, and overdrive is a car racing game provided by the company.

Use of AI-Enabled Features in Retail

iv. **Apple***
v. **Banjo**: It is a company that uses AI technology to process the different information and real-time events taking place on social media platforms to provide the desired results to its partners. The platform was developed in 2013 after the Boston Marathon Bombing to use social media and effective decision-making better.
vi. **CloudMinds**: The company is responsible for making the robots that human beings can control. It is developing a product that they call it as "cloud intelligence-based systems" for robots. The technology is believed to combine the machine with humans rather than treating it as a separate entity.
vii. **Facebook***
viii. **Google***
ix. **H2O**: H2O claims to be "the world's leading open-source deep learning platform." Data scientists worldwide use the platform, and more than 10,000 organizations are believed to use this worldwide. The company's products include its H2O platform, the Sparkling Water framework that combines H2O and Spark, the Steam AI Engine for developers, and Driverless AI, which promises "AI to do AI."
x. **IBM**: IBM is in AI since the 1950s and is believed to be a leader since then. The flagship AI project of IBM is Watson. It uses the AI capabilities such as natural language processing and machine learning to reveal insights from unstructured data.
xi. **Intel**: Intel has invested a lot in AI technologies and considers it essential for the organization. The company has acquired numerous of AI-based companies and invested in different start-ups.
xii. **ViSenze**: ViSenze's uses AI technology to recommend similar products to the online buyers. The company uses a computer vision algorithm along with machine learning technology to process the millions of images. The platform uses visual sensing to suggest the best possible match with the products' price and availability.

*Details of the following companies are given in Section 4.6.1

4.6 AI AND RETAIL

Knowingly or unknowingly, we all interact with AI technology in our day-to-day life. Whether we make a Google search or buy a product on Amazon, we all use AI, from hiring a cab to selecting a flight. Whenever a consumer goes online, he leaves his digital footprints. They make choices, preferences, share locations, and buying behavior, which generate enormous data. The data may be generated through emails, social media activities, smartphones, and other electronic devices connected through the internet. According to the salesforce fifth state of marketing report (2018), the median number of data sources may jump 50% within 2 years, from 10 in 2017 to 15 in 2019. However, these vast amounts of customer data are impossible to be analyzed and utilized by using traditional methods and human minds. To come out with

How AI Platforms Create Value

PLATFORM PROVIDES BRAND
- Virtual shelf space
- A single channel for marketing, sales, and service
- Data on consumer preferences, purchases, and media exposure
- Payment for goods sold
- Product fulfillment
- A trust halo

BRAND PROVIDES PLATFORM
- Listing and promotional fees
- Product information
- Innovations tailored to consumers' needs
- Knowledge about product category

PLATFORM PROVIDES CONSUMER
- Customized recommendations
- Automated routine purchases
- Convenience and savings
- Reduced complexity
- Continual scanning for better deals

CONSUMER PROVIDES PLATFORM
- Payment for goods
- Information on product preferences, purchases, and use
- Information on price sensitivity, risk tolerance, and privacy expectations
- Loyalty in exchange for good recommendations

SOURCE "MARKETING IN THE AGE OF ALEXA," BY NIRAJ DAWAR, MAY–JUNE 2018   © HBR.ORG

FIGURE 4.6 How AI creates value. (Niraj Dawar, 2018.)

meaningful information and intelligence from the available data within a limited time frame, AI is extensively used by marketers (Lichtenthaler, 2020). These data play a vital role in designing personalized Ad campaigns, content automation, personalized marketing targeting the potential customers, customer relationship management of the organization, and marketing automation through AI system. The use of AI helps organizations increase their productivity, improve customer engagement, increase sales and effective customer services, and create a strong relationship with the customer. The importance of AI can be understood through Forbes Insights 2018 research; 84% of the executives have accepted that AI in marketing is essential for their organization's future and will play a key role in their marketing strategy (Figure 4.6).

For applications, Chitkara et al. (2017) identify three main ways in which AI can contribute to a business:

a. **Assisted Intelligence**: It is considered to be the most basic level of AI. It is primarily used for automating the simple process and amplifying the value of current activity. For example, Gmail uses the technology to sort the daily emails into "Primary," "Social," and "Promotion" default tabs. It is also used for image classification by different organizations.
b. **Augmented Intelligence**: It is the technology that focuses on the assistive role of the AI. It enables any platform or individual to perform a task that would not have been possible without it. Assisted intelligence helps organizations test any decision with a low possible risk using the complex computer realities model. The technology is used effectively in the retail sector, legal research, and personal budgeting.

c. **Autonomous Intelligence**: The technology develops and deploys machines that can act independently without any human interventions. It can alter the nature of the task and change the business processes according to the requirement, for example, the use of autonomous intelligence by YouTube or Netflix to suggest the next video; another example is the WeChat platform by the Tencent company.

4.6.1 Digital Giants and AI

In the modern-day retail sector, a tremendous amount of data is generated through various sources. The humongous amount of data cannot be processed or analyzed using the traditional methods. Therefore, retail companies rely on the latest AI technology to analyze the data and extract meaningful information (Lichtenthaler, 2020). The retail sector is believed to be one of the early adopters of the AI technology. Organizations effectively use the technology for personalized suggestions, image-based product search, voice search, and preference-based browsing. Other uses of the technology are efficient delivery management, inventory management, and customer demand anticipation (NITI Aayog, 2018). According to the IBM 2019 Marketing Trends, 61% of the company executives have accepted that machine learning and AI are most significant for their companies. Companies like Amazon, Alibaba, and Flipkart are using AI to identify the customers' purchase intent through the use proximity model to increase the organization's sales and revenue. Similarly, Google uses an Audience Targeting platform based on AI technology for creating Ad campaigns for targeting the desired audience based on their past online searches, habits, location, and affinity, which makes the campaign more effective for marketers and customers (Figure 4.7).

Some retail organizations have their in house AI platforms, while others use the AI platform of these big companies for different purposes. Various uses of AI on different online platforms have been discussed below.

FIGURE 4.7 Impact of AI on marketing operations. (Kate Leggett, Forester research.)

a. **Amazon**

Amazon has it's in-house AI R&D unit and uses the technology across its platform. It provides its machine learning platform to other business units to predict and find patterns in their data for its various possible uses. The company uses AI for predictive analytics, product recommendation, personalized selling, and various autonomous processes on its online platform. In 2014, the company launched its voice assistant platform named Alexa on its Echo devices. These smart speakers have been endowed with a range of skills; they can respond to various user questions, suggest products, learn over time, and enhance their capabilities using machine learning technology. Alexa can also be integrated with other products like air conditioners, cars, or mobiles.

b. **Apple**

Like Amazon's Alexa, Apple has its voice assistant platform Siri introduced in 2011. Since its inception, the platform has evolved a lot, and the company is using this AI-enabled assistance seamlessly into its devices. Apple has also provided a Siri on-boarding site where all the users can try out different ranges of possibilities. The customers also use the platform for any product or information search using voice commands or controlling other electronic devices. The company is also developing a processor named "Apple Neural Engine," designed especially for AI-related tasks.

c. **Baidu**

Baidu is a company based in China and is one of the leading search engines in its country. Baidu has its own AI-focused lab and dedicated team for the research and development of the technology. The company is reaping its investment rewards to develop improved AI technology such as natural language processing and voice recognition. The company believes its technology to be superior to its western counterparts and provides more refined results. Besides the search engine platform, Baidu offers its partners a range of AI-powered solutions like Baidu Brain and a voice assistant platform (DuerOS).

d. **Facebook**

Facebook is using AI-based solutions and services to improve the user experience and engagement on its platform. The company appointed Professor Yann LeCun, from the University of New York, as its new head of the machine learning division to work on its FAIR "Facebook Artificial Intelligence Research" project. The company also announced that it will invest Euro 10 million into its European facility for the development of AI technology. Currently, the company is focused on language problems like question answering and dynamic memory using AI. It has also developed an image recognition app named as Facebook Picture Search. The platform is also used for providing personalized marketing solutions to its partners.

e. **Google**

Google is a leading search engine around the world and one of the leading investors in AI. The main focus of the Google is on machine learning technologies to improve its text to speech translation, language translations, visual search, and prediction capabilities. The customers often use the voice-enabled AI platform for product search. Similarly, the company's Smart Reply platform drafts and sends automated emails based on the previous responses. Google was one of the first companies that involved its users in enhancing its AI platforms' capabilities.

4.7 AI TOOLS

AI is predicted to play a vital role in driving future retailing growth (Grewal et al., 2017). Retailers have realized the importance of AI in order to enhance the consumer shopping experience. The need is also inspired by the past results achieved by implementing AI technology by the leading global players such as Amazon, Alibaba, and Ikea (Watson et al., 2018; Hoy, 2018). The consumer's most common AI tools for their shopping decisions are chatbots, voice assistants, and augmented reality (Turban et al., 2017, as mentioned in Chopra, 2019). These tools assisted the consumers in information search (Brandtzaeg and Følstad, 2017) and enhanced the user experience (Chen and Tsai, 2012). Companies are using AI technologies such as natural language processing (NLP), speech recognition, and image recognition for digital assistant recommendations contextual online advertising chatbots in the retail sector. Various uses of AI technology in the retail sector have been discussed in detail below:

4.7.1 Chatbots

Alan Turing, along with Joseph Weizenbaum, worked on developing a system that can communicate like humans. That was the 1950s when the concept of a chatbot was thought of for the first time. Later, the first chatbot program was invented, named Eliza. "A chatbot is a computer program that simulates human conversation or chat through AI" (Simon, 2019; Riikkinen et al., 2018). Chatbots consist of a structured query language database that stores questions and answers using AI metadata (Allison, 2012). The AI-enabled chatbots can respond to the users' queries, learn from various experiences using machine learning, and improve accordingly. AI-enabled chatbots are capable of converting any form of available data into conversational information. Interaction with the consumers is stored, processed, and used to learn more about the consumer, leading to more personalized and better services (Riikkinen et al., 2018). AI becomes an essential aspect of the currently available chatbots as it can react to user's input and respond accordingly (Duijst, 2017). The latest AI chatbot technologies include learning word representations (Bojanowski et al., 2016) and text classification (Joulin et al., 2016). The development in the various AI technologies such as NLP has made the present chatbots more effective and leads to broader acceptability in other fields.

Different researchers have divided chatbots into different categories. According to Hung et al. (2009), a chatbot can be divided into system-initiated, user-initiated, and mixed-initiative systems based on conversation. In the system-initiated chatbot, the conversation is initiated by the chatbot, and in the user-initiated chatbot, the user leads the conversation, and a chatbot in which two ways conversation initiation is possible is called mixed-initiative system. The chatbot can further be divided into scripted chatbot and artificial intelligent bots (GSMA Intelligence, 2017). Scripted chatbots or retrieval-based chatbots have limited use of AI. It uses a decision tree to provide a simple response to the users. These bots are not very useful when the task or the query becomes complex. Simultaneously, the artificial intelligence bots or generative chatbots use modern AI technology such as machine learning and NLP to perform the complex tasks. It can answer the contextual questions based on past interactions with the user.

According to Smiers (2017), chatbots can be divided into three maturity levels based on the bots' interaction, intelligence, and integration capabilities (see Figure 4.8). The interaction area depicts that the user experience of chatbot is different from the website experience as the interaction with the bot is done through textual input. The interaction is depicted for the chatbot to analyze and understand the input and provide relevant output. The chatbot's back-end is described by the integration area, where it depicts how well the chatbot is integrated with other websites or services.

Any chatbot is placed in level 1 of the maturity model when the interaction is done between human-to-bot, using a single language and one channel (Smiers, 2017). The chatbot can answer simple questions; there are menu-based options and word-based rules for providing the answers and integration with other platforms through simple links. A chatbot can be placed in level 2 of the maturity model when the interaction is done through multi-channel using Multilanguage. The chatbot has NLP, mood detection, and the ability to understand the context. A chatbot can be placed in level 3 of the maturity model when the bots can interact with different bots, multi-person, using multiple languages. These bots have a high level of intelligence and integration capabilities and can self-learn over time.

Currently, chatbots are used in a wide range of fields such as customer service (MarutiTechlabs, 2017), information retrieval (Shawar and Atwell, 2005), education (Letzter, 2016), business, and e-commerce (Chai et al., 2001). Chatbots are believed

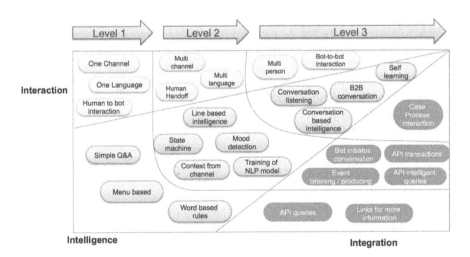

FIGURE 4.8 Maturity model for chatbots. (Smiers, 2017.)

to simplify product search in online shopping (Rowley, 2000). Chatbots can engage consumers in social conversations and are believed to be an effective solution for consumers' routine questions (Vincze, 2017). The benefit of using a chatbot for marketing and customer support service is that they can be made available 24 × 7. They can simultaneously handle several customers; they help provide personalized marketing and customer support services, helping marketers create a more customer-friendly ambiance and increase customer satisfaction. According to the LivePerson survey report (2017), 67% of 5,000 consumers surveyed, located in six different countries, use chatbots for customer support and services. Similarly, the Grand View Research Report (2017) reveals that the global chatbot market may reach $125 billion by 2025. However, chatbots cannot replace human interaction but are very useful in daily business operations for customers and retailers.

4.7.2 PERSONALIZED MARKETING

Every customer wants to be treated individually and expect the service provider to remember them and their needs and preferences, but it becomes nearly impossible for the service provider to keep track of every customer and treat them individually. Here comes the role of AI-based marketing tools, which help marketers make decisions and deploy personalized marketing campaigns. Personalization means, one to one marketing, where companies leverage digital technologies and data analysis to deliver personalized messages and product offerings to consumers. Personalization plays a significant role in brand building, lead generation, customer acquisition, upselling, customer retention, and advocacy in the present marketing world (Salesforce, 2018). In the current digital scenario, every individual can be identified through his email id or IP address, and his navigational behavior can be tracked from his digital footprints. Marketers have access to the customers' real-time data (Kalyanaraman and Sundar, 2006), which targets the individual customer using his demographic characteristics or past buying behavior on online space (Smith et al., 2011).

Online retailers are using AI technology to process this prodigious amount of real-time data to come out with more accurate and highly personalized solutions for the customers (André et al., 2018). The AI-based marketing solutions can help marketers create different content for different users and forward it to the customers at the right time for thousands or millions of customers. The AI systems work on the marketers' predetermined rules and directions to provide the customers' individualized solutions and desired results. This hyper-personalization is based on the predictive behavior of the individual using the machine learning technology. Personalization ultimately leads to more satisfied and loyal customers and increases customer retention (Holland and Baker, 2001).

Figure 4.9, sourced from salesforce research, shows the percentage of marketers who say that personalization improves and helps in the organizations' marketing process. According to it, 54% of respondents have agreed that it plays a significant role in brand building, and 44% have accepted it to be essential for lead generation. Personalization also helps marketers in customer acquisition, upselling, customer retention, and advocacy. AI-based personalization services also increase customer satisfaction by providing more relevant products and services to the consumers.

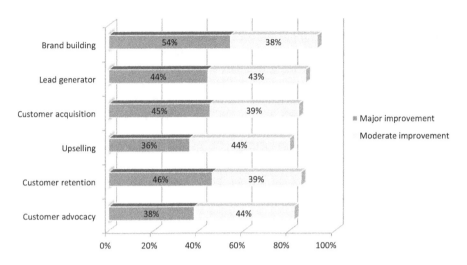

FIGURE 4.9 Percentage of marketers who say personalization improves the following. (Salesforce, 2018.)

According to McTear et al. (2016), the implementation of personalization is constructed in three dimensions:

a. **What to Personalize**: A marketer can personalize the "content, user interface, channel/information access and functionality."
b. **To Whom to Personalize**: After finalizing what to personalize, the next dimension is for whom to personalize. It can either be an individual customer or a group of customers.
c. **Who Executes the Personalization**: There can be either explicit personalization, where input and information is given from the user end or implicit personalization done automatically by the system.

4.7.3 Marketing and Content Automation

Marketing and content automation is concerned with automating content generation and marketing with the use of AI. "Content marketing involves creating, distributing and sharing relevant, compelling and timely content to engage customers at the appropriate point in their buying consideration processes, such that it encourages them to convert to a business building outcome" (Holliman and Rowley, 2014). Much content is generated and needs to be shared with different kinds of customers on social media platforms, web pages, emails, and mobile phones every day. Every individual customer has some different characteristics, need, and preferences to be catered upon. It becomes nearly impossible to collect, analyze, and decide based on the vast amount of data for a large number of customers having different requirements. Online retailers use AI to create a proximity model to put the most relevant content every time they visit the web page. It helps send the relevant email or mobile content to each contact, driving them to the retailer's site. AI uses data to create personalized emails to every company subscriber based on their previous interactions with the brand. AI systems can customize based on what content consumers have consumed, what is on their wish list, what pages they have spent the most time on, and more. AI keeps a close watch on the customers' digital footprints and synchronizes the data produced to the retailer's CRM database. For all these purposes, online retailers widely use AI, as it increases efficiency and reduces the human error and duplicity of message or content involved in the process, while simultaneously improving the content and marketing process's effectiveness. For example, Outbrain and Taboola are AI-based systems used for content recommendation or content management systems by Netflix or Amazon. Such systems propose content that a person is likely to enjoy consuming, given their current choice, allowing them to discover interesting content.

4.7.4 Voice Assistance

Voice assistance is an AI agent that uses NLP, speech synthesis, and voice recognition to process the human voice and provide various solutions (Hoy, 2018). Voice assistance is developing rapidly and can be easily integrated with different platforms and devices like smartphones, PCs, and other gadgets. Various voice assistants available in the market are Google assistance, Alexa and Siri. Apart from these, several companies develop their voice assistance or integrate the available one with their platforms. The present-day voice assistance platforms are different from the earlier voice-activated technologies, which had limited scope. The voice assistance nowadays is based on AI technology, which gives it the ability to perform a variety of tasks such as interpretation and response of text messages, information search, playing music, or making calls as well as controlling the home devices such as air conditioners, fans, and lights (Hoy, 2018). Online retailers are currently using it to provide various solutions such as searching, answering queries, controlling home devices, and delightful experiences. In the context of shopping, voice assistants are assisting consumers with "product information, evaluating product options, and ordering products" (Courtney, 2017). However, these devices' users are concerned about invasion of privacy in homes by these devices and fear leakage of personal information (Alepis and Patsakis, 2017).

Various technologies that are being used in the present-day AI-enabled voice assistants are as follows:
a. Natural Language Processing (NLP)
NLP is the subfield of AI, which is capable of understanding human language. NLP is based on machine learning technology, which gives the voice assistance device the ability to convert human voice into data. "It is this category of AI that will enable computers to understand the hierarchical structure of language and how components of a sentence relate to each other" (Jurafsky and Martin, 2014). The NLP development has opened a wide range of possibilities such as speech to text, grammar correction, modern chatbots, and much more (Kiser, 2016). The recent improvement in NLP is the development in machine learning technology, increased computing power, and linguistic data availability.

b. **Speech Recognition and Voice Recognition**

Speech recognition is the technology that is used to decode and convert the human voice into text. Speech recognition is often confused with voice recognition, which is concerned with identifying the individual's voice.

c. **Natural Language Understanding (NLU)**

The system's ability to understand the meaning of speech and the context in which it is said is achieved by natural language understanding. It is the subset of natural language processing. It is concerned with transforming the human language into a machine-readable format.

d. **Natural Language Generation (NLG)**

Natural language generation gives any system the capability to produce human language output from the various available data input. It allows machine to communicate its output in the form which the humans can understand. As Gartner Research illustrates, "NLP is focused on deriving analytic insights from textual data, NLG is used to synthesize textual content by combining analytic output with contextualized narratives" (Gartner, 2016).

4.7.5 AUGMENTED REALITY

Augmented reality (AR) dates back to the 1950s where it was used for cinematography (Javornik, 2016a). It was the 1990s when the AR started getting more attention from the developer and investors. AR is a technology that interacts with the real-world environment in order to produce augmented images of the environment using computer-generated perceptual information. The automobile industry made the first commercial use of AR in 2008 to create a 3-D simulation. Since then, various kinds of AR apps have emerged in the market, namely, "virtual annotations (Google Glass), virtual try-on, content augmentation, holograms, and projection mapping" (Javornik, 2016b). For AR's operation, a user needs to have a web camera attached to his personal computer or a smartphone (Cabiria, 2012). With various software programs, the images captured with the camera are converted into interactive and augmented images. AI is currently being used with the AR technology for better image processing, with deep learning technology providing more enhanced results.

The AR technology enriches the retail products and environment and will lead to an enhanced consumer experience (Bulearca and Tamarjan, 2010; Huang and Liao, 2017). Some retailers use AR technology for virtual dressing rooms (Yaoyuneyong et al., 2018), where consumers can try the products virtually or try new hairstyles (Magnenat-Thalmann et al., 2006). Similarly, Google fitted with an AR screen provides a virtual experience related to different products (Berryman, 2012). AR gives the consumer the ability to virtually try any product and feel the product's look before actually buying the same. Such technology will enhance the consumer experience and positively respond (Watson et al., 2018). AR-based interactive technology is still in the development stage and is used by very few retailers. While some initial studies believe that it can enhance consumer shopping experience (Huang and Liao, 2017). It can be beneficial for the customers during situations like COVID19, where people would like to interact less with the outside world.

Use of AI-Enabled Features in Retail

4.7.6 Visual Search

Visual search or image recognition is the technology that helps the customer to search for any product using the image. Instead of the traditional text search method, where a consumer needs to know the exact keyword, a visual search can use a smartphone camera or already available image on a pc to get the required information. Some of the example of visual search is Google lens or Amazon's deep lens. It uses AI-powered technology such as deep learning and machine learning to understand these images' content and context. Perhaps the best illustration of what deep neural nets can do is object recognition. It finds the exact match of the images and offers other options such as related articles, price, relevant reviews, and shopping options. Google, Amazon, and Pinterest are some of the leading visual search engines today. The technology is still in its early stage of development, and it may take some time for customers to be accustomed to searching with an image, but recent trends show it is quickly being adopted by retailers (Boyd, 2018) (Figure 4.10).

Various technologies that make the visual search possible are as follows:

a. **Image Processing**

The ability of a system to process and edit any digital image is achieved through image processing technology. Image processing is essential for a system to understand its subject matter. An image is processed either through analog processing or digital processing. The analog processing method is used for processing 2-D images, while the digital processing operates using finite digital data of an image, such as pixels. The present-day AI technology uses digital processing for processing the image.

b. **Object Detection and Image Recognition**

Object detection and image recognition are closely related to two different technologies. Object detection is concerned with identifying a specific object within an image. For example, if we want to know the car's exact location in any picture, the system will create an output image with a car

FIGURE 4.10 Object detection with convolutional neural networks. (Tan and Lim, 2018.)

enclosed in a box with its location. At the same time, image recognition can identify and label the subject matter in any image. Image recognition will create an output with enclosed boxes around the car and label the other objects in the image, such as trees, bikes, and signboards.

c. **Computer Vision**

Computer vision can be defined as "the ability for computers to see imagery through mathematical representations of three-dimensional shape and appearance" (Szeliski, 2011). It gives the system the ability to comprehend the meaning and context of any image similar to humans.

4.8 METHODOLOGY

The chapter's particular section is concerned with the methodology adopted and data analysis to find the features that the customers find significant for them and how satisfied they are with the same. The data required for the study were collected in two stages. In the first stage, personal interview was conducted of the customers, using different platforms for online shopping. The purpose of the interview was to gain insight into the different AI-based features which customers find necessary in any online platform. After analyzing the response in stage 1, visual search, voice assistance, AR, personalization, content automation, and chatbots were identified as essential features. All the features have been discussed in detail in the section above. The identified features are used further in designing a questionnaire, which forms the research instrument in the subsequent empirical research survey.

In the second stage of the study, the data were collected from different customers residing in the Varanasi district of Uttar Pradesh through structured questionnaires. The questionnaire included questions to know the customers' demography and different features significant for customers to be measured on a 5-point Likert Scale. The questionnaire further included questions to measure the satisfaction level of the customers. The sample size was calculated using the often-used formula $n = \dfrac{z^2}{4*(\text{margin of error})^2}$ (Rao, 2013). It comes out to be 97 with a 95% level of confidence with an error margin of 10%, which we round to 100. The data were collected using systematic sampling where every fourth customer receiving the product's delivery was contacted with a local distributor of online products.

4.8.1 DATA ANALYSIS AND RESULTS

In the initial analysis, we found that out of 100 customers from whom the data were collected, 58 were male and 42 were female. All the customers were in the age group of 20–60. The majority of the users were below the age group of 40, as they are the frequent online platforms user. The mean level of the features identified during the first stage was obtained across the 100 observations. After that, the t-value was computed for the customers' features that the customers think are essential for them to. The t-value was associated with the null hypothesis, H_0: Mean ≤ 3 and H_1: Mean > 3. The cut off level of 3 was taken, as it was the median of the scale. The significance level was 5%, so the critical region was > 1.645 for the one-tail test. Thus, whichever feature has a *t*-value > 1.645 was considered significantly crucial by the customers. Table 4.6 summarizes the different mean values and t-values for different features. Out of the six features, customers find personalization, voice assistance, content automation, and chatbots as essential. Personalization was considered the most significant feature of all, with a mean value of 4.22 and a *t*-value of 15.27412; after that, voice search was considered the most crucial AI-based feature with a mean value of 4.04. Simultaneously, visual search and AI-based AR features were not considered significantly crucial by the customers. The mean value of the visual search and AR was calculated as 2.78 and 2.04, respectively, and the *t*-value was also lower than the significant level and hence was considered not very important.

In the case of the satisfaction level, a similar procedure was followed as mentioned above. The mean satisfaction rating was obtained, and their t-value was determined. Table 4.7 summarizes the different mean values and t-values for the satisfaction of different features. Customers were found more satisfied with the voice search as compared to other AI features. The mean value for the voice assistance was computed to be 4.01, and the *t*-value was 14.98238. For personalization, the mean value was computed as 3.21 and the *t*-value as 2.08334. Since only for voice assistance and personalization, the *t*-value was greater than 1.645, it signifies that the online customers were satisfied by these two features only. The customers were not very satisfied

TABLE 4.6
Average Importance Rating of the AI Features by the Online Customers

AI Features	Importance Mean	*t*-Value
Personalization	4.22	15.27412
Voice assistance	4.04	13.15879
Content automation	3.68	7.57589
Chatbot	3.58	6.51926
Visual search	2.78	−3.12201
Augmented reality	2.4	−6.75960

TABLE 4.7
Average Satisfaction Rating of the AI Features by the Online Customers

AI Features	Satisfaction Mean	*t*-Value
Voice assistance	4.01	14.98238
Personalization	3.21	2.08334
Content automation	3.02	0.23139
Chatbot	2.51	−6.36975
Visual search	2.35	−6.78902
Augmented reality	2.53	−7.69738

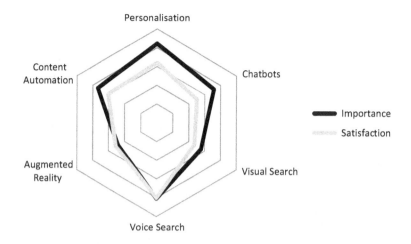

FIGURE 4.11 Difference between the importance and level of satisfaction.

with the other AI features such as chatbot, visual search, and AR with a mean value less than three and a t-value lower than the significance level. For content automation, the mean value was 3.02, and the t-value was 0.23139, which is lower than the level of significance and hence was considered not significant by the customers.

The mean rating on importance and satisfaction is presented diagrammatically in Figure 4.11. The graph shows the gaps between the mean importance level and the mean satisfaction level on all six features. It was observed that there existed a large gap between importance level and satisfaction level on four features. In the case of voice search, the customers were found to be highly satisfied, and in the case of AR, the expectation and satisfaction were both found to be on the lower side. The two results also show that the customers consider personalization the most important but are not satisfied with it then voice search. There was a large gap between the importance and satisfaction level of the customers for personalization. When it comes to chatbots, the customer finds it a significantly important feature as it provides an automated chat facility with 24/7 access to the customer for the different purposes. However, the satisfaction level for the AI-based chatbots is on the lower side and customers seem to be unsatisfied with it.

4.9 CONCLUSION AND FUTURE TRENDS

The fourth industrial revolution, notably the AI, is transforming the AI retail to a great extent. As various researches have been done on its technical aspects, a need was felt to conduct the study from the consumer perspective. The study is a gateway for understanding consumer perception concerning the use of AI tools. An insight into the practice of AI application in the online retail sector is an attempt to contribute to a shared understanding of this phenomenon between involved actors, such as marketers and consumers. The study described AI technology, how it emerged, its components, and its uses in the current retail world. We discussed the different AI technology components such as deep learning and machine learning, their functions, and their current retail sector uses. Though the technology came into existence in the 1950s, it is widely accepted today because of the large amount of data, high computation capabilities, and broader possible use of the technology. The study moves further on identifying the different AI-based features which the consumer finds significant for them. Personalization, chatbots, voice search, content automation, visual search, and AR were found during the first study stage. Later the mean value and the t-value of the feature were computed, and it was found that personalization is considered the most significant feature by the customers, followed by voice search, content automation, and chatbots. The customer considered visual search and AR features as less significant for them. It could be because both these technologies are in their infant stage and still under-developed. Very few retailers and customers are still using the visual search and AR-based features.

Further, the satisfaction level of the consumer for this feature was also obtained. In the satisfaction level, a similar process was adopted as computation of the importance level. The mean value and the t-value were computed for all the significant features identified in stage one. It was observed that the customers find personalization as the most crucial feature but are more satisfied with the voice search feature as compared to others. It was also observed that the customers were unsatisfied with other features such as chatbots, visual search, and AR.

The study, however, has some limitations. The sample size of the study is small and is limited to one city. A large sample size covering a larger geographical area would have been beneficial for the study. Thus, future research may be conducted with a large sample size, including customers from different areas, to improve its generalization. The analysis was done with simple statistical tools such as mean and t-value. The future study may consider the use of other tools to ensure the results to be more valid. The study has a limited scope of identifying different features and computing their importance and satisfaction level. A factor that may affect the importance and the satisfaction level was not considered. Future research may also consider the effect of different factors such as age, gender, income, or technical knowledge on the use of these AI features and their possible effects.

REFERENCES

Aggarwal, A. (2018). Resurgence of AI during 1983–2010. Available at: https://www.kdnuggets.com/2018/02/resurgence-ai-1983-2010.html. Retrieved: December 12, 2019.

Alepis, E. and Patsakis, C. (2017). Monkey says, monkey does—security and privacy on voice assistants. *IEEE Access*, Vol. 5, pp. 17841–17851. doi:10.1109/ACCESS.2017.2747626.

Allison, D. (2012). Chatbots in the library—is it time?. *Library Hi-Tech*, Vol. 30, No. 1, pp. 95–107.

André, Q., Carmon, Z., Wertenbroch, K., Crum, A., Frank, D., Goldstein, W., Huber, J., Van Boven, L., Weber, B. and Yang, H. (2018). Consumer choice and autonomy in the age of artificial intelligence and Big Data. *Customer Needs and Solutions*, Vol. 5, pp. 28–37. doi:10.1007/s40547-017-0085-8.

Berryman, D.R. (2012). Augmented reality—a review. *Medical Reference Services Quarterly*, Vol. 31, No. 2, pp. 212–218.

Bojanowski, P., Grave, E., Joulin, A. and Mikolov, T. (2016). Enriching word vectors with subword information. *Conference Paper to TACL*. Available at: https://arxiv.org/abs/1607.04606. Retrieved: December 2, 2019.

Boyd, C. (2018). Visual search trends statistics tips and uses in everyday life. https://clark-boyd.medium.com/visual-search-trends-statistics-tips-and-uses-in-everyday-life-d20084dc4b0a. Retrieved: November 4, 2020.

Brandtzaeg, P.B. and Følstad, A. (2017). Why people use chatbots. *International Conference on Internet Science*, Springer, Cham, pp. 377–392.

Brynjolfsson, E., Rock, D. and Syverson, C. (2017). Artificial intelligence and the modern productivity paradox—a clash of expectations and statistics. Available at: https://www.nber.org/chapters/c14007.pdf. Retrieved: December 24, 2019.

Bulearca, M. and Tamarjan, D. (2010). Augmented reality—a sustainable marketing tool?. *Global Business and Management Research—An International Journal*. Vol. 2, Nos. 2–3, pp. 237–252.

Businesswire. (2018). Worldwide spending on cognitive and artificial intelligence systems. Available at: https://www.businesswire.com/news/home/20180322005847/en/Worldwide-Spending-Cognitive-Artificial-Intelligence-Systems-Grow. Retrieved: December 24, 2019.

Cabiria, J. (2012). Augmenting engagement—augmented reality in education. In Wankel, C. and Blessinger, P. (Eds). *Increasing Student Engagement and Retention Using Immersive Interfaces—Virtual Worlds, Gaming, and Simulation*. Emerald Group Publishing, Bingley, pp. 225–251.

Castro, D. and New, J. (2016). The promise of artificial intelligence—70 real-world examples. Available at: https://www2.datainnovation.org/2016-promise-of-ai.pdf. Retrieved: December 27, 2019.

Chai, J.Y., Budzikowska, M., Horvath, V., Nicolov, N., Kambhatla, N. and Zadrozny, W. (2001). Natural language sales assistant—a web-based dialog system for online sales. *Proceedings of the 13th Innovative Applications of Artificial Intelligence Conference IAAI*, Seattle, WA, pp. 19–26.

Chen, C.M. and Tsai, Y.N. (2012). Interactive augmented reality system for enhancing library instruction in elementary schools. *Computers & Education*, Vol. 59, No. 2, pp. 638–652.

Chitkara, R., Rao, A. and Yaung, D. (2017). Leveraging the upcoming disruptions from AI and IoT. *PwC*. Available at: https://www.pwc.com/gx/en/industries/communications/assets/pwc-ai-and-iot.pdf. Retrieved: August 14, 2019.

Chopra, K. (2019). Indian shopper motivation to use artificial intelligence—generating Vroom's expectancy theory of motivation using grounded theory approach. *International Journal of Retail & Distribution Management*, Vol. 47, No. 3, pp. 331–347. doi:10.1108/IJRDM-11-2018-0251.

Courtney, M. (2017). Careless talk costs privacy (censorship digital assistants). *Engineering &Technology*, Vol. 12, No. 10, pp. 50–53.

Datamation. (2017). Top 20 artificial intelligence companies. Available at: https://www.datamation.com/applications/top-20-artificial-intelligence-companies.html. Retrieved: October 15, 2020.

Duijst, D. (2017). Can we improve the user experience of chatbots with personalisation? Thesis submitted to the University of Amsterdam. doi:10.13140/RG.2.2.36112.92165.

Forbes. (2018). AI plus human intelligence is the future of work. https://www.forbes.com/sites/jeannemeister/2018/01/11/ai-plus-human-intelligence-is-the-future-of-work/#6173e97c2bba. Retrieved: January 14, 2020.

Forbes. (2019). Forbes report on AI. https://www.forbes.com/sites/insights-intelai/2019/05/22/welcome-from-forbes-to-a-special-exploration-of-ai-issue-6/?sh=f93f4554650e. Retrieved: January 14, 2020.

Gartner. (2016). Natural language understanding. Available at: https://www.gartner.com/en/information-technology/glossary-/nlu-natural-language-understanding. Retrieved: June 18, 2020.

Grand view. (2017). Global chatbot market. Available at: https://www.grandviewresearch.com/press-release/global-chatbot-market2017. Retrieved: June 12, 2020.

Grewal, D., Roggeveen, A.L. and Nordfält, J. (2017). The future of retailing. *Journal of Retailing*, Vol. 93, No. 1, pp. 1–6.

GSMA. (2018). The mobile economy 2018. Available at: https://www.gsmaintelligence.com/research/?file=061ad2d2417d6ed1ab002da0dbc9ce22&download. Retrieved: June 15, 2020.

GSMA Intelligence. (2017). Chatbots-conversational commerce. Available at: https://www.ecommercewiki.org/topics/119/chatbots-conversational-commerce. Retrieved: October 11, 2020.

Hirschberg, J. and Manning, C.D. (2015). Advances in natural language processing. *Science*, Vol. 349, No. 6245, pp. 261–266. Available at: https://pubmed.ncbi.nlm.nih.gov/26185244/; doi:10.1126/science.aaa8685. Retrieved: October 15, 2020.

Holland, J. and Baker, S.M. (2001). Customer participation in creating site brand loyalty. *Journal of Interactive Marketing*, Vol. 15, No. 4, pp. 34–45.

Holliman, G. and Rowley, J. (2014). Business to business digital content marketing—marketers' perceptions of best practice. *Journal of Research in Interactive Marketing*, Vol. 8, No. 4, pp. 269–293. doi.10.1108/JRIM-02-2014-0013.

Hoy, M. (2018). Alexa, Siri, Cortana, and more—an introduction to voice assistants. *Medical Reference Services Quarterly*, Vol. 37, pp. 81–88. doi:10.1080/02763869.2018.1404391.

Huang, T. and Liao, S. (2017). Creating e-shopping multisensory flow experience through augmented reality interactive technology. *Internet Research*, Vol. 27, No. 2, pp. 449–475.

Hung, V., Elvir, M., Gonzalez, A. and DeMara, R. (2009). Towards a method for evaluating naturalness in conversational dialog systems. *IEEE International Conference on Systems, Man and Cybernetics*, IEEE, San Antonio, TX, p. 1236.

IBM. (2019). Marketing trends report. Available at: https://newsroom.ibm.com/2018-12-17-IBM-Watson-Marketing-Releases-2019-Marketing-Trends-Report-Focused-on-Emerging-Trends-Redefining-the-Profession-in-the-Shift-to-AI. Retrieved: October 11, 2019.

IDC. (2016). Available at: https://www.businesswire.com/news/home/20170925005077/en/IDC-Spending-Guide-Forecasts-Worldwide-Spending-on-Cognitive-and-Artificial-Intelligence-Systems-to-Reach-57.6-Billion-in-2021 Retrieved: July 14, 2019.

Javornik, A. (2016a). Augmented reality—research agenda for studying the impact of its media characteristics on consumer behaviour. *Journal of Retailing and Consumer Services*, Vol. 30, pp. 252–261.

Javornik, A. (2016b). It's an illusion, but it looks real! Consumer affective, cognitive and behavioral responses to augmented reality applications. *Journal of Marketing Management*, Vol. 32, Nos. 9–10, pp. 987–1011.

Joulin, A., Grave, E., Bojanowski, P. and Mikolov, T. (2016). Bag of tricks for efficient text classification. Available at: https://arxiv.org/abs/1607.01759. Retrieved: September 23, 2019.

Jurafsky, D. and Martin, J.H. (2008). *Speech and Language Processing—An Introduction to Natural Language Processing. Computational Linguistics and Speech Recognition.* Prentice-Hall, Upper Saddle River, NJ.

Jurafsky, D. and Martin, J.H. (2014). *Speech and Language Processing—An Introduction to Natural Language Processing. Computational Linguistics and Speech Recognition.* Dorling Kindersley Pvt, Noida.

Kalyanaraman, S. and Sundar, S.S. (2006). The psychological appeal of personalized content in web portals—does customization affect attitudes and behavior?. *Journal of Communication*, Vol. 56, No. 1, pp. 110–132.

Kaplan, A. and Haenlein, M. (2019). Siri, Siri in my hand, who's the fairest in the land? On the interpretations, illustrations and implications of artificial intelligence. *Business Horizons*, Vol. 62, No. 1, pp. 15–25.

Letzter, R. (2016). IBM's brilliant AI just helped teach a grad-level college course. Available at: http://uk.businessinsider.com/watson-ai-became-a-teaching-assistant-2016-5?international=truer=UK&IR=T. Retrieved: June 24, 2017.

Lichtenthaler, U. (2020). Beyond artificial intelligence—why companies need to go the extra step. *Journal of Business Strategy*, Vol. 41, No. 1, pp. 19–26. doi:10.1108/JBS-05-2018-0086.

Liveperson. (2017). Bots in customer care. https://www.liveperson.com/resources/reports/bots-in-customer-care/. Retrieved: October 24, 2020.

Magnenat-Thalmann, N., Montagnol, M., Gupta, R. and Volino, P. (2006). Interactive virtual hair-dressing room. *Computer-Aided Design and Applications*, Vol. 3, No. 5, pp. 535–545.

MarutiTechlabs. (2017). Customer supports bot. Available at: https://www.facebook.com/Customer-Support-Bot-1857341381220252/. Retrieved June 24, 2017.

McCarthy, J., Minsky, M.L., Rochester, N. and Shannon, C.E. (1955). A proposal for the Dartmouth summer research project on artificial intelligence. http://www-formal.stanford.edu/jmc/history/dartmouth/dartmouth.html. Retrieved: June 17, 2019.

McKinsey Global Institute (MGI). (2017a). Artificial intelligence—the next digital frontier?. Available at: https://www.mckinsey.com/~/media/mckinsey/industries/advanced%20electronics/our%20insights/how%20artificial%20intelligence%20can%20deliver%20real%20value%20to%20companies/mgi-artificial-intelligence-discussion-paper.ashx Retrieved: October 24, 2020.

McKinsey Global Institute (MGI). (2017b). Artificial intelligence—implications for China. Available at: https://www.mckinsey.com/~/media/mckinsey/featured%20insights/China/Artificial%20intelligence%20Implications%20for%20China/MGI-Artificial-intelligence-implications-for-China.ashx Retrieved: October 24, 2020.

McTear, M., Callejas, Z. and Griol, D. (2016). *The Conversational Interface*. Springer, Cham.

Minsky, M. (1974). *A Framework for Representing Knowledge*. MIT, Cambridge, MA.

NSTC. (2016). Preparing for the future of artificial intelligence. Executive Office of the President. Available at: https://info.publicintelligence.net/WhiteHouse-ArtificialIntelligencePreparations.pdf. Retrieved: December 4, 2019.

Niraj Dawar. (2018). Marketing in the age of Alexa. Available at: https://hbr.org/2018/05/marketing-in-the-age-of-alexa#. Retrieved: September 14, 2020.

NITI Aayog. (2018). Discussion paper, National Strategy for Artificial Intelligence. https://niti.gov.in/writereaddata/files/document_publication/NationalStrategy-for-AI-Discussion-Paper.pdf. Retrieved: September 14, 2020.

Rao, A. (2017). A strategist's guide to artificial intelligence, Outlook 2017–21, 5. Available at: https://www.strategy-business.com/article/A-Strategists-Guide-to-Artificial-Intelligence Retrieved: August 28, 2019.

Rao, P.H. (2013). *Business Analytics—An Application Focus*. PHI Press, New Delhi.

Riikkinen, M., Saarijärvi, H., Sarlin, P. and Lähteenmäki, I. (2018). Using artificial intelligence to create value in insurance. *International Journal of Bank Marketing*, Vol. 36, No. 6, pp. 1145–1168. doi:10.1108/IJBM-01-2017-0015.

Robert Allen. (2017). Available at: https://www.linkedin.com/pulse/15-applications-artificial-intelligence-marketing-robert-allen/Retrieved: December 4, 2019.

Rowley, J. (2000). Product searching with shopping bots. *Internet Research*, Vol. 10, No. 3, pp. 203–214.

Russell, S. and Norvig, P. (1995). A modern approach. *Artificial Intelligence*, Vol. 25, No. 27, pp. 79–80.

Salesforce. (2018). The 5th state of marketing report. https://www.salesforce.com/blog/2018/12/introducing-fifth-state-of-marketing-report. Retrieved: December 20, 2019.

Salesforce. (2020). The 6th state of marketing report uncovers trends to navigate change. https://www.salesforce.com/blog/top-marketing-trends-navigate-change/. Retrieved: November 14, 2020.

Samuel, A. (1959). Some studies in machine learning using the game of checkers. IBM Journal of Research and Development, Vol. 3, No. 3, pp. 210–229.

Shawar, B. and Atwell, E. (2005). Using corpora in machine-learning chatbot systems. *International Journal of Corpus Linguistics*, Vol. 68, No. 4, pp. 489–516.

Simon, J.-P. (2019). Artificial intelligence—scope, players, markets and geography. *Digital Policy, Regulation and Governance.* doi:10.1108/DPRG-08-2018-0039.

Smiers, L. (2017). How can chatbots meet expectations? Introducing the bot maturity model. Available at: https://www.capgemini.com/2017/04/how-can-chatbots-meet-expectations-introducing-the-bot-maturity//. Retrieved: December 24, 2020.

Smith, J.H., Dinev T. and Xu, H. (2011). Information privacy research—an interdisciplinary review. *MIS Quarterly*, Vol. 35, No. 4, pp. 989–1015.

Stone, P., Brooks, R., Brynjolfsson, E., Calo, R., Etzioni, O., Hager, G., Hirschberg, J., Kalyanakrishnan, S., Kamar, E., Kraus, S., Leyton-Brown, K., Parkes, D., Press, W., Saxenian, A.L., Shah, J., Tambe, M. and Teller, A. (2016). *Artificial Intelligence and Life in 2030. One Hundred Year Study on Artificial Intelligence—Report of the 2015–2016 Study Panel.* Stanford University, Stanford, CA. Available at: http://ai100.stanford.edu/2016-report. Retrieved: July 24, 2020.

Szeliski, R. (2011). *Computer Vision-Algorithms and Applications.* Springer. doi:10.1007/978-1-84882-935-01.

Tan, K.-H. and Lim, B.P. (2018). The artificial intelligence renaissance—deep learning and the road to human level machine intelligence. *APSIPA Transactions on Signal and Information Processing*, Vol. 7, p. e6. doi:10.1017/ATSIP.2018.6.

Turban, E., Outland, J., King, D., Lee, J.K., Liang, T.P. and Turban, D.C. (2017). *Electronic Commerce 2018—A Managerial and Social Networks Perspective.* Springer, Cham.

Van Eeuwen, M. and Van Der Kaap, H. (2017). Mobile conversational commerce—messenger chatbots as the next interface between businesses and consumers. http://essay.utwente.nl/71706/1/vanEeuwen_MA_BMS.pdf. Retrieved: December 14, 2019.

Varian, H. (2018). Artificial intelligence, economics, and industrial organization. NBER Working Paper No. 24839. doi:10.3386/w24839.

Vincze, J. (2017). Virtual reference librarians (Chatbots). *Library Hi Tech News*, Vol. 34 No. 4, pp. 5–8.

Watson, A., Alexander, B. and Salavati, L. (2018). The impact of experiential augmented reality applications on fashion purchase intention. *International Journal of Retail and Distribution Management.* doi:10.1108/IJRDM-06-2017-0117.

Wharton. (2017). Why AI is the new electricity. Available at: https://knowledge.wharton.upenn.edu/article/ai-new-electricity/
Retrieved: October 2, 2020.

Wierenga, B., Casillas, J. and Martínez-López, F.J. (Eds.). (2010). *Marketing Intelligence Systems.* STUDFUZZ 258, pp. 1–8. Springer-Verlag, Berlin, Heidelberg.

Yaoyuneyong, G.S., Pollitte, W.A., Foster, J.K. and Flynn, L.R. (2018). Virtual dressing room media, buying intention and mediation. *Journal of Research in Interactive Marketing.* Vol. 12, No. 1, pp. 125–144.

5 Effective Integration of Lean Operations and Industry 4.0
A Conceptual Overview

Aaron Ratcliffe
Appalachian State University

Maneesh Kumar
Cardiff University

Sriram Narayanan
Michigan State University

CONTENTS

5.1	Introduction	98
5.2	Overview of Lean Principles and Industry 4.0 Tools	100
5.3	Purpose: How Does I4.0 Enhance the Ways Organizations Create Value for Long-Term Competitive Advantage? TW #1; L5P #1	103
5.4	Processes: How Does I4.0 Help Design and Improve Right Processes?	105
	5.4.1 Just-In-Time (JIT) Pillar. TW #2–4; L5P #2–3	105
	5.4.2 Jidoka Pillar. TW #2–4; L5P #2–3	106
	5.4.3 Standardization and Visual Control as the Foundation. TW #6–7; L5P #2–4	107
	5.4.4 Importance of Technology and Its Integration with People and Processes (L5P #8)	107
5.5	People: How Does I4.0 Help Empower and Develop Employees? TW #9–11	108
5.6	Problem-Solving: How Does I4.0 Enhance Problem-Solving, Continuous Improvement, and Organization Learning?	110
5.7	Conclusion	111
References		112

5.1 INTRODUCTION

Organizations face novel challenges and opportunities as they adapt to the shifting competitive landscape brought on by globalization and digitization. Markets now evolve faster. Products rapidly become commodities. Organizations compete as supply chains. Customers demand greater flexibility and responsiveness – enhanced customization, swift product development, integrated product-service combinations, and fast delivery options across multiple channels. Managing more with less is the MANTRA. There is a greater emphasis on corporate social responsibility, accountability to stakeholders, and the impact of today's actions on future generations. Organizations also seek new ways of working and adapting their organizational and supply chain structure to cope with demand variations, supply uncertainty, technological advancements, and end-to-end connectivity requirements in the supply chain. They are under ever-greater pressure to do the right things and do them well: high quality, low cost, fast delivery, high flexibility, customer satisfaction, and positive or zero environmental impact. All these need to be undertaken with fewer resources.

How can one organization perform well in all these dimensions? This question is at the heart of both Lean manufacturing and Industry 4.0. As a management system, Lean examines the question from an organization's perspective by presenting a philosophy oriented toward waste elimination through principles of just-in-time production, intelligent automation, continuous improvement, and standardized processes (Bortolotti et al., 2015). Industry 4.0, the fourth industrial revolution, examines the question from a technological perspective by considering the momentous impact that modern smart technologies, advanced analytics, and end-to-end connectivity across the value chain have on markets and industries (Frank et al., 2019). Lean principles offer a means for maintaining focus on operational effectiveness in the era of rapid technological advancement accelerated by Industry 4.0. New Industry 4.0 capabilities offer a means for enhancing the strength and expanding the scope of Lean practices through value chain integration, automation, and digitization (Chiarini & Kumar, 2020). Lean helps improve productivity and other performance metrics by reducing waste existing in business processes and its supply chain using bundles of soft and hard practices including just-in-time manufacturing, set-up, pull, flow, total productive maintenance (TPM), long-term strategic relationships with suppliers and customers, and statistical process control (Bortolotti et al., 2015; Shah & Ward, 2007; Tortorella et al., 2019). By applying both Lean and Industry 4.0 together, organizations can find ways to increase productivity and stretch the boundaries of the efficient frontier.

Integration of Lean methodologies with Industry 4.0 technologies seems natural, given that they work toward similar objectives, but guidance on how to do this is limited. Together, the two should lead to fruitful solutions that help companies stay competitive, adapt to new challenges, and improve operational performance through reduced complexity, increased flexibility, and improved productivity (Buer et al., 2018; Chiarini & Kumar, 2020; Tortorella et al., 2020). However, there is limited practical evidence on *how to enable integration between the two* and whether *Lean should be the foundation to enable a successful transition to the implementation of Industry 4.0 technologies.* Early indications from the literature show positive outcomes for organizations when integrating Lean with Industry 4.0 adoption instead of a standalone approach (Chiarini & Kumar, 2020; Kolberg et al., 2017; Sanders et al., 2016; Tortorella et al., 2019, 2020). Literature suggests that Lean is more useful for the first round of improvement and streamlining the processes before automating or digitizing the process using Industry 4.0 technologies (Chiarini & Kumar, 2020). This conclusion is also based on limited evidence of the sequence of integration of the two approaches, making it difficult to conclude if Lean should be the foundation for Industry 4.0 implementation. Does the conclusion imply that their introduction should be sequentially ordered, with first a strong foundation in Lean, or are there cases where their synergies emerge through implementation in a synchronous or cyclical manner? How can one characterize the relationship between Lean and Industry 4.0? Is it unidirectional or bidirectional? Ultimately, important questions remain.

A key challenge with Lean is that as its methodologies become more widespread across organizations, the potential impacts are diluted. The complete shift in organizational culture and management philosophy required to reap the full benefits of Lean may be too daunting or risky for many organizations, yet the tools themselves are accessible and ubiquitous. A common metaphor for The Toyota Production System (TPS) is that of a house – remove the foundation or one of the key pillars and the house will fall. Yet, organizations tend to go for the low-hanging fruit and cherry-pick only the most accessible practices on an ad-hoc basis. Seminars and workshops on Lean and Six Sigma and certain terms and tools such as PDCA, "just-in-time", 5S, value stream maps, and kaizen have become a commonplace in manufacturing and service organizations. But this does not mean that every organization has seen as dramatic shift in operational performance as Toyota. The same can be said for emerging Industry 4.0 technologies such as AI, machine learning, and smart products. The technologies are in vogue and becoming more accessible for consumers and supply chains – but without a coherent strategy, their impact may fall flat. For this reason, an organization must keep in mind a guiding philosophy and principles upon which to build a strategy for capturing the synergies found when combining the various elements. Implementing Lean principles before the adoption of new technology helps to eliminate waste and streamline processes – the new digital technology cannot alone fix a broken process – but more importantly, it provides a guiding philosophy so that organizations stay focused on the long-range goals and navigate with a true north in mind.

The COVID-19 pandemic and an increase in other global and natural disasters have brought an inflexion point for Industry 4.0. The global health and economic crisis forced organizations to take extraordinary measures to protect their people and ensure continuity of operations. This resulted in accelerated innovation in organizations and the emergence of new business models for remote working and remote management of operations and supply chain activities. There were a sudden demand surge for Industry 4.0 technologies such as additive manufacturing (AM) and technology for PPE and ventilator manufacturing, increased robotics and automation to help implement social distancing guidelines in production settings, Cloud services for remote working, monitoring, and management (e.g., Track and Trace app for remote monitoring and management of COVID19 patients), AI and machine learning technology for enabling the use of chatbots in the service industry, and advanced robotics for managing COVID patients in hospital wards. All these applications have greater alignment with Lean philosophy as the technology has enabled organizations to streamline processes and practices and allow their employees to work safely without compromising efficiency. The 3D printing has cut down the number of steps in manufacturing a mask or a visor; advanced robotics creates capacity for clinicians to focus on more important jobs; chatbots drop the unwanted non-value-added calls that occupy the customer service time in a call center environment; augmented reality enabled through wearable glasses accelerates training and knowledge transfer compared to the traditional training method.

This chapter discusses how to employ Lean principles and augment Lean practices by adopting Industry 4.0 technologies. Our goal is not only to find overlaps between Lean and Industry 4.0 but to (1) characterize the synergistic relationship between the two, (2) showcase real-world examples of how to successfully utilize them together, and (3) illustrate how the reach of Lean practices can be extended throughout the value chain with end-to-end integration enabled by Industry 4.0. In Section 5.2, we provide a brief overview of the five Lean principles (Womack & Jones, 1996) and the 14 principles of the 4Ps model of The Toyota Way (Liker, 2004) and key Industry 4.0 (I4.0) technologies. We build upon those integrative frameworks, denoting the five Lean principles as L5P and The Toyota Way as TW, to structure the body of the chapter around the 4Ps to examine how Industry 4.0 enhances the application of foundational principles of Lean systems.

   I. **Purpose and Value**: How does I4.0 enhance the way customer organizations create value for long-term competitive advantage? (TW #1; L5P #1);
  II. **Processes**: How does I4.0 help design and improve the right processes? (TW #2–8; L5P #2–4);
 III. **People**: How does I4.0 help empower and develop employees and value chain partners? (TW #9–11)
  IV. **Problem-Solving and Perfection:** How does I4.0 enhance problem-solving, continuous improvement, and organization learning? (TW #12–14; L5P #5)

Throughout the sections, we emphasize how I4.0 helps expand the scope of Lean initiatives across organizational boundaries to the value chain and ecosystem levels through increased connectivity, visibility, and end-to-end integration.

5.2 OVERVIEW OF LEAN PRINCIPLES AND INDUSTRY 4.0 TOOLS

The five Lean principles (L5P) described by Womack & Jones (1996) offer a core framework for capturing the essence of Lean:

   1. **Identify Value**: Specify value from the standpoint of the end customer by product family.
   2. **Map the Value Stream**: Identify all the steps in the value stream for each product family, eliminating whenever possible those steps that do not create value.

3. **Create Flow**: Make the value-creating steps occur in tight sequence so the product will flow smoothly toward the customer.
4. **Establish Pull**: As flow is introduced, let customers pull value from the next upstream activity.
5. **Seek Perfection**: As value is specified, value streams are identified, wasted steps are removed, and flow and pull are introduced, begin the process again and continue until a state of perfection is reached in which the perfect value is created with no waste (Figure 5.1).

An alternative framework is the 4Ps model of The Toyota Way (Liker, 2004). Using these two frameworks as a guide, we will structure our discussion of the integration between Lean practices and Industry 4.0 into the four key sections defined above (Figure 5.2).

At its core, Lean focuses on the elimination of waste. Ohno defined seven types of TIMWOOD wastes (i.e., transportation, inventory, motion, waiting, overproduction, overprocessing, and defect) across the organizational processes and its supply

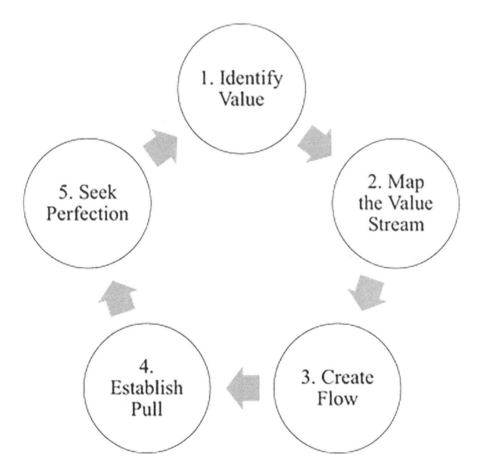

FIGURE 5.1 Five Lean principles. (Adapted from Womack & Jones, 1996.)

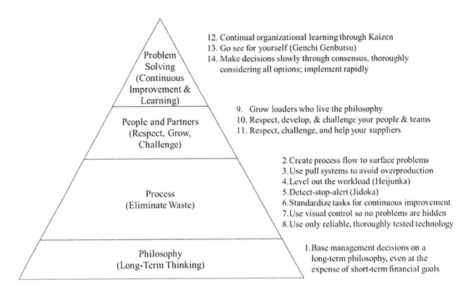

FIGURE 5.2 P model of The Toyota Way. (Adapted from Liker, 2004.)

chain by identifying those activities that do not add value and eliminate or simplify those activities, thereby streamlining the process and creating standardized routines (Womack & Jones, 1996). The streamlining and simplification of processes are facilitated through the collection of Lean tools including 5S, value stream mapping (VSM), 5S, Just-In-Time (JIT), Heijunka, Jidoka, Kanban, Poka-Yoke and Single Minute Exchange of Die (SMED), and TPM. A summary of the Lean toolbox and its purpose is provided in Table 5.1.

TABLE 5.1
Lean Tools and Purpose

Lean Tools	Focus or Purpose
Value Stream Mapping (VSM)	Visual map showing information and material flow, and wastes in the current state of the process and envisioning a less wasteful future state map
Just-In-Time (JIT)	Only producing when customer trigger an order; this requires continuous flow from raw materials to finished products and synchronizing processes
Heijunka	Leveling orders to minimize waste, avoid building inventory, and balancing processes according to the takt-time or rhythm of orders
5S	Housekeeping tool for organizing spaces to perform work efficiently, effectively, and safely; foundation for visual factory, help in improving material, and information flow

(Continued)

TABLE 5.1 (*Continued*)
Lean Tools and Purpose

Lean Tools	Focus or Purpose
Cellular Manufacturing	Objective to establish one-piece flow by designing a specific U-shaped layout where workstations are in a sequence and employees multi-tasking that supports a leveled flow of materials with minimal transport or delay
Single Minute Exchange of Die (SMED)	System to significantly reduce set-up time for equipment changeover
Kanban	Based on JIT/pull systems principle, a visual card for communication between upstream and downstream processes for triggering the production of components from upstream workstations only when it is needed and in the right quantities according to the takt-time of received orders
Total Productive Maintenance	Continuous improvement of equipment effectiveness by reducing machine stoppages and failures through preventive and predictive maintenance carried out by workers (autonomous maintenance) and professionals
Poka-Yoke, mistake proofing	Control mechanisms within process to alert operators before error occurs; device or sensors that prevent the occurrence of mistakes or defects
Jidoka – autonomation	Implementing automatic systems of detecting problems in machineries but letting the worker the possibility of solving the problem stopping the machine

Source: Adapted from Chiarini and Kumar (2020).

Industry 4.0 connects the real world with virtual reality using a combination of technologies including cyber physical system (CPS), cloud computing, Internet of Things (IoT), big data and analytics, robotics, and other smart technologies (Frank et al., 2019; Srinivasan et al., 2020; Tortorella et al., 2019). Industry 4.0 is being branded as a strategic model for dramatically improving metrics such as cost, productivity, quality, customer satisfaction and lead time by end-to-end integration of the supply chain through automation and digitization (Buer et al., 2018; Kolberg et al., 2017). These advanced information and manufacturing technologies allow the vertical integration of manufacturing processes and horizontal and end-to-end integration of the supply chain to transform and upgrade current industrial manufacturing and service organizations (Chiarini & Kumar, 2020).

5.3 PURPOSE: HOW DOES I4.0 ENHANCE THE WAYS ORGANIZATIONS CREATE VALUE FOR LONG-TERM COMPETITIVE ADVANTAGE? TW #1; L5P #1

Industry 4.0 signifies a disruptive shift in the way that organizations compete and do business. Integration of Industry 4.0 is often a capital-intensive effort where long-term benefits justify the high upfront costs, training, and other time investments in the short-term. But I4.0 connects to philosophy in a deeper way, as organizations are left considering "What business am I in?" (Porter & Heppelmann, 2014). The traditional means of creating value for customers are shifting under I4.0. There is an increasing emphasis on value cocreation with customers as we move to digital ecosystems (Li et al., 2020). Examples of this include John Deere moving away from selling tractors to offering smart farming services, Volvo shift from selling trucks to selling uptime, Rolls-Royce shifting from selling engine to flying time in the air, and Siemens Logistics offering maintenance solutions to airports instead of selling technology.

As organizations tackle Industry 4.0, they build competitive advantage through better and new means of value creation. A central principle of Lean is waste elimination. While elimination of waste creates value, Lean and Industry 4.0 methods need to better create value for specific processes or value streams. Value streams in many firms cut across functional disciplines and can help in identifying new products and services. Development of a value stream requires four different steps that involve the (1) identification of the process and the value-adding elements, (2) establishing a current state of the process, (3) designing the future state, and (4) consolidating the changes to the future state (Tortorella et al., 2020). However, it is important to understand that firms are a collection of multiple value streams that clearly provide end-customer value. A central tenet of Lean is the importance of making the internal firm process more customer-focused. While several use cases exist of different value streams – for example, IoT technologies in predictive maintenance, developing an effective understanding of integrating the diverse value streams for competitive advantage is a key aspect of Lean-Industry 4.0 integration. In this regard, Industry 4.0 initiatives that are more customer-focused are likely to take advantage of changing the fundamental processes that deliver customers value and provide additional opportunities to create future value for the firm.

For example, firms like Volvo AB that have invested in connectivity products are fundamentally changing their supply chains by altering service processes and redesigning the delivery channels for these services. Production of connected vehicles in this regard comes with a new infrastructure of a call center that enables effective response to incidents, creation of a dealer infrastructure and preparedness that allows the organization to become more customer facing, and possibly allowing factories to better assess products on the road directly, implementing design changes. Thus, an effective competitive advantage for such firms does not just come from implementing technology initiatives within a product or a process but a holistic change to the approach the firm takes. Another example of a large firm that focused on IoT enabling a heating, ventilation, and air conditioning system firm also resulted in substantial changes to not only the product, but the supply chain of how the product was produced, delivered, used, and maintained. As Industry 4.0 technologies provide additional process-level visibility, individual and collective value streams need more focus for organizations to create a long-run competitive advantage.

In each of the above example, the focus has shifted from operational efficiency to operational excellence by working jointly with customers to understand their long-term value and accordingly making changes in their offerings with the aid of Industry

Lean Operations and Industry 4.0   105

4.0 technologies integrated with their products and services. These companies have developed innovative solution ecosystems that guarantee long-term product performance, i.e., creating long-term value for the customer. Industry 4.0 technologies, including smart sensors, have created the opportunity for manufacturers and service providers to understand their customers' usage pattern and use those data to further improve the product and service offerings to existing and new customers.

5.4 PROCESSES: HOW DOES I4.0 HELP DESIGN AND IMPROVE RIGHT PROCESSES?

The TPS is often described as a house with the two key pillars of Just-In-Time production and Jidoka (intelligent automation). These pillars stand on a foundation of standardization, continuous improvement, and visual control. We discuss how I4.0 enhances each pillar and the foundation separately.

5.4.1 JUST-IN-TIME (JIT) PILLAR. TW #2–4; L5P #2–3.

The JIT pillar of the TPS exhorts having the right part in the right amount at the right place at the right time. It relies on the key principles of continuous flow (L5P #3 and TW #2), pull systems (L5P #4 and TW #3), and demand leveling (TW #4). These principles help to eliminate waste and force problem-solving through limited work-in-process inventory. Industry 4.0 enhances application of these principles through end-to-end integration and connectivity that enables greater visibility and synchronization of material flows and new flexible production technologies.

Continuous Flow (L5P #3, TW #2) seeks to achieve a constant flow of product in small batches using line balancing and set-up time reduction (SMED) techniques (Womack & Jones, 1996). The adage is to "run like the tortoise not like the hare." Batch production leads to high levels of work-in-process inventory that "hide the rocks" or the process problems below the surface. Continuous steady flow reinforces doing things right the first time. Small batch sizes expose problems and force immediate problem-solving, and flexibility improves as lead times for individual units decrease.

Pull systems (L5P #4, TW #3) maintain minimum inventory and only produce when a customer triggers an order (Ohno, 1988). Pull systems are often executed using Kanban or pull signals between upstream and downstream processes that authorize work at the upstream process only when it is needed and in the right quantities according to the takt-time of received orders. Pull systems help to synchronize production, eliminate waste, and force problem-solving. Accurate inventory data are critical in a pull system which operates with minimal buffer and safety stocks (Holweg, 2007). To implement a pull system across the supply chain requires accurate and timely information sharing and synchronization across stages since any delay or disruption of material flows may delay production and lead to failure in meeting customer deadlines.

Demand Leveling (Heijunka) (TW #4): focuses on creating an even workflow while allowing for mixed model production that matches the customer demand and balances processes according to the takt-time or rhythm of orders.

IoT, Cyber Physical Systems, and cloud computing allow organizations to adopt a digitalized supply chain that allows for real-time tracking of material flows, accurate inventory data, and information sharing across the supply chain: e.g., e-Kanban (MacKerron et al., 2014). Smart sensors have been combined with Kanban systems to reduce inventory and overcome Lean implementation barriers at Würth Industries (Kolberg et al., 2017; Sanders et al., 2016). Denkena et al. (2014) derive key performance indicators to improve material flow from shop floor data autonomously collected using RFID tags. AM offers flexible equipment to facilitate continuous flow, pull systems, and set-up time reduction (Chen & Lin, 2017; Mayr et al., 2018). 3D Printers can be quickly reconfigured digitally for mix-model production (Heijunka) and customization or to replace broken parts and reduce MRO inventory (Chen & Lin, 2017; Sanders et al., 2016).

At AB Volvo, IoT allows for better inventory planning at the dealers. IoT sensors alert service centers of fault codes on incoming trucks so that they can triage and diagnose repair issues in advance and ensure that they have the correct inventory of parts available. Less inventory is wasted trying to anticipate every possible problem that may arrive.

5.4.2 Jidoka Pillar. TW #2–4; L5P #2–3

Jidoka, the second pillar of TPS, emphasizes exposing problems and solving them as they occur. Detect-Stop-Alert exemplifies how humans and machines work together on monitoring and production functions as part of "automation with a human touch". "Built-in quality" mechanisms on machines are responsible for automatically identifying defects in the process, stopping production to avoid further defects, and alerting the worker (Ohno, 1988). This liberates operators to focus their skills on high labor-content tasks as opposed to constant monitoring. Workers are also responsible for solving problems immediately, identifying root causes, mistake-proofing, and installing counter measures. Jidoka shows the alignment with information and communication technologies to address waste and defects in the manufacturing processes, though employees still play a central role in controlling those wastes (Monden, 2011). Mistake-proofing tool is now seen used in day-to-day activities in manufacturing, services, and public sector organizations for identifying abnormalities (e.g., color coding, alarm or sound to alert workers, auto-cut of supply to prevent error).

Jidoka practices are inbuilt within Industry 4.0 technologies that allow real-time monitoring of processes and instigate automatic signals when products or processes are not in a state of control. Ma et al. (2017) created a CPS-based smart Jidoka for the piston rod assembly of an automotive company and revealed that the production system leads to an improved flexibility, nearly 50% reduced costs, and shorter lead time. The use of machine learning, big data, and predictive analytics capabilities augment the practice of TPM within the workplace, or it can be said that the most modern machines will have the inbuilt feature of predictive maintenance, i.e., the machine generating warning signal before it breaks down (Srinivasan et al., 2020).

To develop Airport 4.0 solutions for maintenance of the baggage handling system at the Heathrow airport, Koenig et al. (2019) used Industry 4.0 technologies including sensors on wheels of cart carrying customer's baggage. Cloud computing and using machine learning to generate a visual dashboard enabled real-time monitoring of the condition of the wheels by maintenance staff. The dashboard allowed the maintenance team to only act when they find data linked to vibration of the wheel beyond the threshold limit, which is an indication that the wheel needs to be replaced. This condition monitoring solution not only reduced the cost of maintenance and penalty cost due to delayed baggage (which miss the flight due to breakdown of the cart carrying luggage), but it also helped the maintenance team to move from reactive to more proactive maintenance approach enabled by Industry 4.0 technologies. From 2016 to 2018, when the error-proof predictive maintenance solution was operationalized, there was not a single breakdown of cart in the Heathrow tunnel carrying more than 2,500 carts between terminal buildings.

5.4.3 STANDARDIZATION AND VISUAL CONTROL AS THE FOUNDATION. TW #6–7; L5P #2–4

Most of the Lean tools reinforce the principle of visual control and transparency, making it easier to visualize waste or identify problems in the process (Buer et al., 2018). This is aligned with Industry 4.0 technologies that are considered to improve the product or process quality of manufacturing or service organizations (Li et al., 2020; Tortorella & Fettermann, 2018). The digital technologies listed in Table 5.2 improve traceability, visibility, and connectivity between the organization's operations and its supply chain, hence improving the decision-making process (Kolberg et al., 2017). The 5S tool for standard housekeeping and ensuring everything is in its place can be made more effective when combined with sensors and technologies. In an example shared by Chiarini and Kumar (2020), a manufacturing organization in Italy attached a smart sensor to one of its expensive tools that were often misplaced by operators. In times of emergency, it was not easy to trace who is using the tool. The embedded sensor helped improve traceability by knowing the exact location and the person using the tool. Similar IoT solution has been applied in hospital settings to trace the wheelchairs that get left by porters in a multi-story hospital building which are sometimes difficult to trace or their "whereabouts unknown" (Stevens, 2019). Stevens (2019) suggested that more than 25% of the wheelchairs in the US hospitals are lost or stolen each year. The IoT solutions were applied to valuable medical equipment in the US healthcare setting for improving the efficiency and effectiveness for hospital operations, avoiding serious consequences linked to admitting and discharging patients due to lack of traceability of the medical equipment, and providing timely care to the patients.

5.4.4 IMPORTANCE OF TECHNOLOGY AND ITS INTEGRATION WITH PEOPLE AND PROCESSES (L5P #8)

As organizations bring in Industry 4.0 technologies into their shop floor, it is important that these tools focus on both tools and process that are usable and do not remain at the stage where such use cases are "idle." For example, Denso implemented I4.0 technologies that focused on implementing technologies that make it easy for operators to manage processes. Furthermore, a key element of Industry 4.0 technologies

TABLE 5.2
List of Industry 4.0 Technologies and Their Purpose

- **IoT and Cyber Physical Systems**: Network to connect anything with the Internet through information sensing equipment to conduct information exchange and communications in order to achieve smart recognitions, positioning, tracing, analysis, etc.
- **Big Data and Analytics**: Big data is characterized by an immense volume, variety and velocity of data across a wide range of networks. Analytics have evolved from business intelligence and decision support systems enabling organizations to analyze big data to support evidence-based decision making and action taking
- **Artificial Intelligence (AI) and Machine Learning**: Artificial intelligence concerns the ability of machines to carry out tasks typically performed by a human intelligence. Machine learning is a branch of AI where machines have access to data and learn by themselves, making decisions or predictions
- **Cloud Technologies**: Cloud computing involves delivering hosted services over the Internet. These services are typically infrastructure-as-a-service, platform-as-a-service, and software-as-a-service
- **3D Printing, Additive Manufacturing**: 3D printing, included in the broader term of AM, refers to the various processes used in the manufacture of products, by depositing or fusing materials layer by layer
- **Digital Automation with Sensors and Smart Sensors**: Machines and manufacturing processes embedded with sensors capable of collecting data, measuring, analyzing, and triggering other processes
- **Collaborative and Autonomous Mobile Robots (COBOT and AMR)**: Collaborative robot (COBOT) is a robot intended to physically interact with worker. The robot could be restricted in a shared workplace or be able to move itself autonomously
- **AR and SHI**: Communication systems which allow people to interact with a number of smart technologies such as screens, 3D glasses, exoskeletons, etc. which augment human abilities

Source: Adapted from Chiarini and Kumar (2020).

is that these technologies are organic. Specifically, they should not eliminate individuals (mechanistic), rather they remain useful to eliminate unpredictability in the human element of production. In that sense, making I4.0 technologies more people-friendly makes them more relevant to implementing traditional production with a human touch. We discuss more on the potential for Industry 4.0 technologies to facilitate the human touch in Section 5.5 below.

5.5 PEOPLE: HOW DOES I4.0 HELP EMPOWER AND DEVELOP EMPLOYEES? TW #9–11

Employee development is an integral aspect of I4.0 and its connection to technology. With regard to Industry 4.0, multiple aspects of employee development come to the fore. First is that I4.0 is blurring the difference between white and blue-collar workforce. Second, I4.0 is also helping teams weave through information and help integrate individuals with different abilities into operational processes.

In examples shared before regarding predictive maintenance at Heathrow airport (Koenig et al., 2019), the role of maintenance team changes completely from reactive mode of maintenance (i.e., run to break or planned maintenance) to more proactive mode based on developing predictive maintenance solutions using smart sensors, cloud, and AI. In their new role, they are expected to understand the visual boards and take actions when they identify any anomaly in the real-time vibration data generated from 2500 carts inside Heathrow tunnel. The operators were upskilled to learn how to interpret the data and take necessary actions at the right time. So, apart from technical skills, now maintenance operators are also expected to demonstrate analytical and decision-making skill-sets. This example itself shows how Industry 4.0 technologies are going to affect the roles of maintenance employees in future.

Artificial intelligence and machine learning are changing the way customers interact with service providers in the era of Industry 4.0. Most of the standard queries in the call center environment are now being automated and replaced by chatbots, with customer service representative upskilled and reskilled to deal with more complex queries of the customer. The idea here is that problems, queries, or operations that are standardized and do not require customization can be automated and digitized using AI and ML, and more value-added jobs can be tasked to human operators. Warehousing operations are seeing increasing application of wearable technologies such as augmented reality (AR) glasses that enables humans to access digital information through a layer of information positioned on top of the physical world. The picking, sorting, and loading operations in the warehouse does not require a high level of intelligence (Chiarini & Kumar, 2020). Thus organization does not need to recruit high-skilled employees for picking and loading operations; the skilled-employees can be deployed elsewhere in the operations which requires more analytical and problem-solving skills (Srinivasan et al., 2020).

As the boundary between blue-collar and white-collar workers are getting blurred in the era of Industry 4.0, it forces organizations to rethink their human resource management (HRM) strategy. The HR function needs to work closely with operations and other functions to understand the skill-sets required to embrace Industry 4.0 solutions, upskilling and reskilling of employees based on the new requirements to use Industry 4.0 technologies, changing the recruitment strategy and the criteria to recruit employees not only based on their technical skills but also their propensity to learn new skills. As the Industry 4.0 technology list keeps growing, organizations that have recruited employees based on their learning style will adapt quickly as it is easier to learn new skills if your employees have intrinsic motivation to change and to learn.

Furthermore, I4.0 technologies have also enabled individuals with disabilities engage in operational processes (Narayaran et al., 2019). A key element of employment of individuals with disabilities in operational settings is the variability in potential performance outcomes. With new technologies such as accessible drawings, power industrial exoskeletons, wearables that track employee activity and heart-rate with a view to keeping them healthy, intelligent personal assistants that act as aid in performing tasks that otherwise would be harder to perform (Romero et al., 2016). Telerobotic systems aid in providing physical support and strength for employees (Auquilla et al., 2019). For example, research is being undertaken on using AR technology for training employees with specific instructions to complete specific jobs (Funk et al., 2016). As new technologies and processes are envisioned, it is important to consider how process changes, individual capabilities, and technology choices interact (Narayanan & Terris, 2020; Narayaran et al., 2019). Specifically, effective process design and experimentation is a key element of such technology implementations (Narayanan & Terris, 2020). Overall, with regard to empowering and developing people, integration of Lean and Industry 4.0 provides substantial opportunities.

5.6 PROBLEM-SOLVING: HOW DOES I4.0 ENHANCE PROBLEM-SOLVING, CONTINUOUS IMPROVEMENT, AND ORGANIZATION LEARNING?

Continuous improvement (kaizen) is a critical to pursuit of perfection within Lean systems as seen in Lean principle #5 and The Fourth P of The Toyota Way Framework. Kaizen is often implemented by applying the scientific method to process improvement through Plan-Do-Study-Act (PDSA) or Shewhart Deming Cycles. These cycles have been adapted within Six Sigma as Define-Measure-Analyze-Improve-Control (DMAIC) cycles to incorporate a fifth step dedicated to controlling the improved process for variability. These iterative cycles of improvement can be applied rapidly to narrow the scope and address specific problems within a broader cycle of organizational improvement, e.g. a kaizen event (PDSA cycle) within the "Improve" phase of DMAIC. However, kaizen more importantly represents an organizational mindset of continually striving to be better. We now consider how Industry 4.0 enhances Lean principles within each stage of the improvement cycle, with a broad principle of kaizen in mind.

Within the Define (Plan) Phase, the organization frames the problem, brainstorms potential solutions, and develops a plan by setting its key goals and metrics for the improvement cycle, considering the available data, and developing hypotheses and predictions. I4.0 helps widen the scope of the improvement cycle plans across organizational boundaries of the value chain through increased connectivity, visibility, and end-to-end integration. I4.0 strengthens idea generation during the Plan Stage through improved crowd-sourcing, communications, and cloud computing technologies that facilitate people working in teams to creatively propose innovative solutions to problems. Team members working across geographical boundaries can all apply the "see for yourself" principle (Genchi Genbutsu) through cloud computing, virtual reality and AR and Smart Human Interfaces (SHI). Physical presence in the space is still critical for human observation and trust building among workers; however, virtual gemba walks save on travel costs and time for remote quality managers and have a necessity for resilience in response to the COVID-19 pandemic which has limited travel and enforced social distancing guidelines within production and service facilities.

During the Measure (Do) phase, the organization carries out the plan, maps and documents the process, and collects the necessary data. I4.0 boosts this phase through enhanced data collection and process visualization. Big data, IoT, and smart sensors offer new sources of data for defining and collecting data for key metrics.

During the Analyze (check) phase, the organization investigates the results, examines the data against the hypotheses, and identifies root causes of variability and poor performance in the process. I4.0 magnifies these abilities through advanced analytics. Machine learning and artificial intelligence offer new means for finding relationships and causes of variability in the data. As with the planning stage, cloud computing and enhanced communication technologies make it easier for cross-functional teams to collaborate on interpreting the results and what was learned from the experiment conducted in the earlier phase. Cloud computing also makes advanced analytical tools more widely accessible to decision-makers across levels and functions of the organization and value chain.

During the Improve (Act) Phase, the organization moves forward on the basis of what was learned by implementing the plan on a wider scale, rejecting the plan, or repeating the improvement cycle with further knowledge of how to address and eliminate the root causes of the problem. I4.0 technologies augment capabilities to roll out improvements quickly. Updates to computer systems can be made quickly through 5G networks. Robots can be reprogrammed to work differently more quickly than humans. Virtual reality makes it possible for employees at one facility to "see for themselves" the best practices, processes or working conditions of another. Overall, I4.0 speeds up the kaizen cycle process by allowing ideas to be generated, data to be collected and analyzed, and plans to be implemented much more rapidly than before.

During the Control phase, the organization monitors behavior under the new process to control for variability. Refer back to Section 5.4.3. Process for information on how I4.0 enhances control of variability.

5.7 CONCLUSION

By reviewing the emerging literature on Lean and Industry 4.0 and examining specific industry examples, we discussed how I4.0 enhances the application of Lean principles along four key themes of purpose, process, people, and problem-solving. I4.0 is disrupting business models and the way organizations create value and achieve competitive advantage as customers look for integrated combinations of physical and digital products and services and supply chains move toward product-service systems and digital ecosystems. Through increased connectivity, visibility, end-to-end integration, and flexible production, I4.0 widens the scope over which processes can be designed and standardized to eliminate waste across organizational boundaries by applying the pillars of just-in-time production and intelligent automation (jidoka). I4.0 supports respect, growth, and empowerment of employees and partners by enhancing human physical abilities, facilitating better communication and teamwork across geographical boundaries, and valuing workers as knowledge workers free from repetitive manual labor to creatively solve problems, design better processes, and work with customers to cocreate value. I4.0 augments problem-solving capabilities through enhanced idea generation, consensus building, data collection, and advanced analytics.

Within the context of I4.0, it is more important than ever that managers have a full and clear understanding of how various Lean principles work together as part of an overall management system and philosophy. Ad-hoc adoption of Lean principles and I4.0 technologies based on industry bandwagons and buzzwords may cause their joint benefits to be diluted and their synergies to be blocked. Organizations must be able to take risks and evolve their strategies, value chains, and processes as quickly as customers and markets are evolving around them. However, organizations must deliberate in their approach so that new technologies serve the people and the process (TW Principle #8) and are aligned with organizational values and strategic. As with technology in the past, change management will be crucial for building consensus and considering all options while working quickly to implement solutions (TW Principle #13). But the time to leverage these opportunities is now – for the sake of building a more sustainable future economy and thriving in the ecosystems of tomorrow.

REFERENCES

Auquilla, A. R., Salamea, H. T., Alvarado-Cando, O., Molina, J. K., & Cedillo, P. A. S. (2019, 15–18 Oct. 2019). Implementation of a Telerobotic System Based on the Kinect Sensor for the Inclusion of People with Physical Disabilities in the Industrial Sector. Paper presented at the *2019 IEEE 4th Colombian Conference on Automatic Control (CCAC)*. IEEE, Medellin, Colombia.

Bortolotti, T., Boscari, S., & Danese, P. (2015). Successful lean implementation—Organizational culture and soft lean practices. *International Journal of Production Economics, 160*, 182–201. doi:10.1016/j.ijpe.2014.10.013.

Buer, S.-V., Strandhagen, J. O., & Chan, F. T. S. (2018). The link between Industry 4.0 and lean manufacturing—Mapping current research and establishing a research agenda. *International Journal of Production Research, 56*(8), 2924–2940. doi:10.1080/00207543.2018.1442945.

Chen, T., & Lin, Y.-C. (2017). Feasibility evaluation and optimization of a smart manufacturing system based on 3D printing—A review. *International Journal of Intelligent Systems, 32*(4), 394–413. doi:10.1002/int.21866.

Chiarini, A., & Kumar, M. (2020). Lean Six Sigma and Industry 4.0 integration for operational excellence—Evidence from Italian manufacturing companies. *Production Planning & Control*, 1–18. doi:10.1080/09537287.2020.1784485.

Denkena, B., Dengler, B., Doreth, K., Krull, C., & Horton, G. (2014). Interpretation and optimization of material flow via system behavior reconstruction. *Production Engineering, 8*(5), 659–668. doi:10.1007/s11740-014-0545-z.

Frank, A. G., Dalenogare, L. S., & Ayala, N. F. (2019). Industry 4.0 technologies—Implementation patterns in manufacturing companies. *International Journal of Production Economics, 210*, 15–26. doi:10.1016/j.ijpe.2019.01.004.

Funk, M., Kosch, T., Kettner, R., Korn, O., & Schmidt, A. (2016). motionEAP—An overview of 4 years of combining industrial assembly with augmented reality for Industry 4.0. In *Proceedings of the 16th International Conference on Knowledge Technologies and Data-Driven Business (I-Know 2016)* (p. 4), Graz, Austria.

Holweg, M. (2007). The genealogy of lean production. *Journal of Operations Management, 25*(2), 420–437. doi:10.1016/j.jom.2006.04.001.

Koenig, F., Found, P. A., & Kumar, M. (2019). Innovative airport 4.0 condition-based maintenance system for baggage handling DCV systems. *International Journal of Productivity and Performance Management, 68*(3), 561–577. doi:10.1108/IJPPM-04-2018-0136.

Kolberg, D., Knobloch, J., & Zühlke, D. (2017). Towards a lean automation interface for workstations. *International Journal of Production Research, 55*(10), 2845–2856. doi:10.1080/00207543.2016.1223384.

Li, A. Q., Kumar, M., Claes, B., & Found, P. (2020). The state-of-the-art of the theory on product-service systems. *International Journal of Production Economics*, 222, 107491. doi:10.1016/j.ijpe.2019.09.012.

Liker, J. K. (2004). *Toyota Way—14 Management Principles from the World's Greatest Manufacturer*. New York: McGraw-Hill Education.

Ma, J., Wang, Q., & Zhao, Z. (2017). SLAE–CPS—Smart Lean automation engine enabled by cyber-physical systems technologies. *Sensors, 17*(7), 1500.

MacKerron, G., Kumar, M., Kumar, V., & Esain, A. (2014). Supplier replenishment policy using e-Kanban—A framework for successful implementation. *Production Planning & Control*, 25(2), 161–175. doi:10.1080/09537287.2013.782950.

Mayr, A., Weigelt, M., Kühl, A., Grimm, S., Erll, A., Potzel, M., & Franke, J. (2018). Lean 4.0—A conceptual conjunction of lean management and Industry 4.0. *Procedia CIRP, 72*, 622–628. doi:10.1016/j.procir.2018.03.292.

Monden, Y. (2011). *Toyota Production System—An Integrated Approach to Just-In-Time*, 4th Edition. Boca Raton, FL: CRC Press.

Narayanan, S., & Terris, E. (2020). Inclusive manufacturing—The impact of disability diversity on productivity in a work integration social enterprise. *Manufacturing & Service Operations Management, 22*(6), 1112–1130. doi:10.1287/msom.2020.0940.

Narayaran, S., Terris, E., & Sharma, A. (2019). Abilities-first—Steps to create a human-centric inclusive supply chain. *Supply Chain Management Review* (November/December), 34–41.

Ohno, T. (1988). *Toyota Production System—Beyond Large-Scale Production*. Boca Raton, FL: CRC Press.

Porter, M. E., & Heppelmann, J. E. (2014). How smart, connected products are transforming competition. *Harvard Business Review, 92*(11), 64–88.

Romero, L. E., Chatterjee, P., & Armentano, R. L. (2016). An IoT approach for integration of computational intelligence and wearable sensors for Parkinson's disease diagnosis and monitoring. *Health and Technology, 6*(3), 167–172. doi:10.1007/s12553-016-0148-0.

Sanders, A., Elangeswaran, C., & Wulfsberg, J. P. (2016). Industry 4.0 implies lean manufacturing—Research activities in Industry 4.0 function as enablers for lean manufacturing. *Journal of Industrial Engineering and Management (JIEM), 9*(3), 811–833.

Shah, R., & Ward, P. T. (2007). Defining and developing measures of lean production. *Journal of Operations Management, 25*(4), 785–805. doi:10.1016/j.jom.2007.01.019.

Srinivasan, R., Kumar, M., & Narayanan, S. (2020). Human resource management in an Industry 4.0 era—A supply chain management perspective. In T. Y. Choi, J. J. Li, D. S. Rogers, T. Schoenherr, & S. M. Wagner (Eds.), *The Oxford Handbook of Supply Chain Management*. Oxford University Press. doi:10.1093/oxfordhb/9780190066727.013.39.

Stevens, L. M. (2019, October 28). Where's the wheelchair? IoT allows tracking valuable medical equipment, *Lucidworks*. Accessed April 17, 2021, from https://lucidworks.com/post/iot-healthcare-tracking-medical-equipment/.

Tortorella, G. L., & Fettermann, D. (2018). Implementation of Industry 4.0 and lean production in Brazilian manufacturing companies. *International Journal of Production Research, 56*(8), 2975–2987. doi:10.1080/00207543.2017.1391420.

Tortorella, G. L., Giglio, R., & van Dun, D. H. (2019). Industry 4.0 adoption as a moderator of the impact of lean production practices on operational performance improvement. *International Journal of Operations & Production Management, 39*(6/7/8), 860–886. doi:10.1108/IJOPM-01-2019-0005.

Tortorella, G. L., Pradhan, N., Macias de Anda, E., Trevino Martinez, S., Sawhney, R., & Kumar, M. (2020). Designing lean value streams in the fourth industrial revolution era—Proposition of technology-integrated guidelines. *International Journal of Production Research, 58*(16), 5020–5033. doi:10.1080/00207543.2020.1743893.

Womack, J. P., & Jones, D. T. (1996). *Lean Thinking*, 1st Edition. New York: Simon and Schuster.

6 Opportunities and Risks
Use of Autonomous Vehicles in Logistics

Nikunj S. Yagnik
CVM University

CONTENTS

6.1 Introduction 115
6.2 Introduction to AVs: Strategies and Materials... 117
    6.2.1 Historical Development 117
    6.2.2 Technology.. 118
    6.2.3 Legislation and Liability.. 119
    6.2.4 Values.. 120
    6.2.5 Human Engineering.. 120
6.3 Autonomous Vehicle in Logistics: Utilization and Outcome 120
    6.3.1 AVs in Indoor Logistics: Applications.. 121
    6.3.2 AVs in Outdoor Logistics: Applications 122
    6.3.3 Autonomous Vehicles in Long-Haul Freight Transport................... 123
    6.3.4 Autonomous Vehicles in Highway Trucking.................................. 123
    6.3.5 Contemplation on Logistics Operation... 124
6.4 Autonomous Vehicles in Logistics: Risks 124
6.5 Conclusion 125
References... 125

6.1 INTRODUCTION

Currently, many researchers are working on advanced technologies in the field of autonomous cooperation to be used in logistics and supply chain. Many advantages and positive outcomes are known to this point (e.g., flexibility and complex system lustiness) (Carlo et.al. 2014; Casey 2014) though apart from numerous positive effects associated with intelligent cooperation technologies, possible negative impacts through applying such systems in routine logistics environment have to be compelled to be discussed and resolved. These potential adverse influences would possibly persuade new or have an effect on prevailing dangers in provision like nondeterminism of the underlying system, the bullwhip effect, etc.

Autonomous vehicles (AVs) have emerged amidst fast technological advancements and have attracted abundant attention, thanks to their potential to remodel mobility. At current rates of development, we can estimate that companies reminiscent of Tesla, Volvo, GM, Audi, Mercedes, and Nissan can unleash AVs available by 2025, and moreover we do predicts that 25% of the world market will be occupied by AVs by 2040. However, it's unclear whether or not the total extent of AVs' edges will be realized, and there are considerations concerning the new risks that will emerge. Governance ways will facilitate to maximize AVs' potential edges, minimizing the risks that usually result from adopting new technologies.

AVs will be categorized in step with their degree of automation.

The SAE known as the Society of Automotive Engineers well defined the basic five levels of automation widely adopted across the major global organizations. The level one is known as assisted automation where dynamic driving tasks performed by driver and a couple of partial automations are available. At level three (conditional automation), the driving force is predicted to manage the vehicle sometimes upon warnings. At level four, automation is being considered as high automation in which a vehicle is assessed as totally autonomous, and level five comes under full automation, where it is expected that the vehicle should operate without driver properly under all possible environmental conditions In our study, we have a tendency to specialize in interconnected and total AVs as they represent a larger elementary shift in society. Consequently, the subsequent question must be replied: How will the operation and employment of intelligent cooperative technological systems touch the associated dangers in logistic systems?

The presented paper itself is a possible answer to the given question. This paper addressed the potential risks associated in terms of autonomous cooperation technologies to be used and implemented in the area of logistics and supply chain.

As result, expressive objective to be represented in this paper which includes development of autonomous assistance along with overall classification to supply-chain risks. The associated degree of analytical goal then seeks to depict possible consequences between the introduction of autonomous collaboration technologies and, as a result, the categories of supply-chain risks. In conclusion, the praxeological goal is to generate hypotheses about the interrelationships among the application of autonomous collaboration technologies and their effect to an establishment's risks as a springboard to further research. The move to completely independent transport is an advancement through truck units on thruways and little urban conveyance units. Associated cars organize and test the advances for the inevitable progressive driverless experience. The concept of self-driving, independent vehicles has been talked around for a long time. Whether from the car division, science fiction, or enormous information devotees, the appearance of cars, trucks, and buses that explore and drive themselves has been a common desire. The reality in terms of fully automated vehicles to be placed at several field of supply chain is getting progressively closer and over the following decade, numerous anticipate to see a few significant progresses presented at scale in a few parts of the world, in spite of the fact that at distinctive speeds in several divisions and in numerous locales.

This paper briefly and crisply provides strategies and materials used in the current development of AVs logistics.

6.2 INTRODUCTION TO AVs: STRATEGIES AND MATERIALS

The concept of AVs became more practical over a period of time. The existing literature addressing a wide range of AV-related issues started to emerge. The majority of those research areas have no clear connection to supply. Nevertheless, once making an attempt to make connection with supply, however AVs may have an effect on supplying and be enforced in observe, it's necessary to bear in mind of the analysis tired these completely different fields. Thus, we tend to provide a temporary summary of the foremost common analysis topic regarding AVs that might be of interest for supplying purpose of reading. Because a comprehensive examination of each of those disciplines of research is beyond the scope of this paper, a brief explanation is provided along with some reading recommendations.

6.2.1 Historical Development

Dreams of AVs and robotized interstates within the mid-20th century remained to a great extent within the attention of futurists and fans of science fiction. Disney, for example, revealed a program called "Magic Thruway USA" in 1958, which imagined a future with, among other things, automated vehicles powered by colored interstate paths and working with addresses coded on punch cards. Until the mid-1980s, the fundamental computer and other developments required to implement (and reconsider) these ideals were not widely available. The progresses made within the final 25 years of long time can be caught in terms of three progressive waves of developmental picks up.

A second goal of the study was to create semi-autonomous and autonomous vehicles that relied on highway infrastructure only infrequently, if at all. A team led by Ernst Dickmanns at the Bundeswehr College Munich in Germany produced a vision-guided vehicle that could travel at speeds of up to 100 km/h while remaining immobile in the early 1980s (Lantos Mâarton, 2011). NavLab 1 to NavLab 11 were a series of vehicles developed by Carnegie Mellon University's NavLab from the mid-1980s to the early 2000s. In July 1995, on a "No Hands Over America" visit, NavLab 5 traveled across the country, managing itself 98 percent of the time even when a human administrator was present. In July 1995, during a "No Hands Over America" visit, NavLab 5 drove across the country, with the vehicle controlling itself 98% of the time and a human administrator controlling the throttle and brakes the remaining 2%. Other similar projects around the world were looking for ways to build and process early AV and thruway ideas. In the early 2000s, the Park-Shuttle, an automated public road transportation system, was launched in the Netherlands (Shladover, 2007; Panatoya, 2003; Andréasson, 2001). The US government has also began developing self-driving vehicles, primarily for military use. Demo I (US Armed Forces), Demo II (DARPA), and Demo III (US Armed Forces) were all backed by the US government (Hong, 2000). Demo III demonstrated how autonomous ground vehicles can traverse miles of challenging off-road terrain while dodging trees and rocks (2001). A real-time control system, which could be a multileveled control system, was offered by the National Institute for Measurement and Technology. Individual vehicle control (e.g., throttle, controlling, and braking) was combined with high-level goals to stimulate the manufacturing of large groups of vehicles (Bellutta, 2000; Shoemaker, 1998; Hong, 2002). Volkswagen's "Transitory Autopilot" (TAP) system will govern the automobile semi-autonomously at speeds up to 130 km/h. The major issue in terms of high speed driving for autonomous vehicle is accident-free driving, according to Jürgen Leohold, head of Volkswagen Gather. This technology was created as part of the European Union's HAVEit (Highly Computerized Vehicles for Intelligent Transportation) program, which has a budget of $40 million (Flemisch, 2011). It boasts several driver-assist capabilities, including variable cruise control and side observing for safer lane-changing, thanks to a radar framework, laser scanner, and ultrasonic sensors. When the car is in TAP mode, it keeps a safe distance from the vehicle in front of it, checks the lane markers to stay in the middle, and automatically slows down as it approaches a street curve. It helps to avoid accidents caused by inattentive drivers. The driver, on the other hand, retains control and has the ability to override the car's behavior at any time (Bartels, 2014).

6.2.2 Technology

Until now, the great majority of audiovisual research has concentrated on the devices' technological characteristics. Because autonomous vehicle technology is still in its infancy and has only recently begun to be commercialized, it is not yet fully matured. It is currently in the process of being developed and tested. However, most new automobiles come equipped with automation features like as forward collision warning (Dagan et al., 2004;), adaptive control (Vahidi & Eskandarian, 2003), and lane departure warning (Dagan et al., 2004; ; Casey, 2014; Lee, 2002). It's crucial to realize that there is no such thing as a single AV technology while trying to grasp how it works. Rather, the capacity of an autonomous vehicle to drive itself is built on a mixture of hardware and advanced software technologies that work together to organize the device and enable autonomy. It's difficult to present a full picture of the state-of-the-art in AV technologies, as Shanker et al. (2013) point out, because several OEMs are building an AV using their own methodology while keeping it hidden, partially owing to competition considerations. AVs, on the other hand, have four important technologies that allow them to operate autonomously in the complex context of daily traffic. Surround perception and modelling, route planning and decision-making, localization and map building, and motion control are examples of such elements for the design of AVs (Siegwart & Nourbakhsh, 2004). Simply defined, an AV must be able to gather data about its surroundings, understand the data, utilize the interpretation to plan the most basic AV actions conceivable, transform these plans into actionable instructions, and execute the orders (Anderson et al., 2014). According to Anderson et al. (2014), AVs must have extensive backup systems that track the activities of numerous key elements and are ready to navigate to a safe parking lot if one of the first units fails.

These capabilities can be attained in one of two ways (Shanker et al., 2013). V2V and V2I communication systems are extensively used in the main. The notion is that the infrastructure informs the automobile about its surroundings, and the car uses its own light detection and ranging (LIDAR) readings to compare to a map database to discover differences as obstacles to avoid. The low cost of the vehicle is a plus, but the vehicle's poor capacity to respond to unanticipated changes, and as a result, the expense of road infrastructure is a drawback. The second system relies on the car's ability to fully observe and analyze its surroundings rather than on environmental feedback. This system has a higher vehicle cost because of the necessity for a variety of cameras, radars, and sensors, as well as a higher sensitivity to weather. The advantage is the ability to adjust more quickly to climate change and a greater level of infrastructure independence. Silberg and Wallace (2012) claim that combining the two strategies would result in higher levels of protection, autonomy, and self-driving capacity than either strategy could achieve on its own, and they explain how they arrived at their conclusion. The final AV will mostly certainly be a mixture of the two technologies stated. Shanker et al. (2013) describe the numerous hardware apparatuses and their functions in an AV technology framework (camera, radar, LIDAR, sensor, GPS/communication, human machine interface, domain controller, and motion control system) in a simple and easy-to-understand manner. Trible et al. (2014), on the other hand, discuss the challenges that developers continue to encounter in terms of perception, artificial intelligence, and decision-making. Cheng (2011) describes in detail the many approaches to the many components that make up the mechanism that allows an AV to move itself. Veres et al. (2011) conducted a technical analysis of AV decision-making methodologies, while Anderson et al. (2014) summarizes the current state of AV technology, with a focus on telematics and connectivity, because, as they argue, AVs will require certain technologies not only for vehicle to infrastructure and vehicle to vehicle, but also to update maps and applications, as well as to provide passengers with infotainment.

6.2.3 LEGISLATION AND LIABILITY

Legal problems and queries of liability ought to be answered so as to permit for AVs. The necessity for legislative change, as well as the question of responsibility, is usually highlighted as major hurdles to the general deployment of AVs, which all rely on human drivers at the moment. As of nowadays, it's not utterly clear, UN agency can take responsibility if property is damaged, or folks are hurt by AVs. This additionally extremely depends on the particular level of automation, since a user UN agency has no risk to interfere with potential harmful maneuvers of the automobile, can't be only command dependent . On the opposite hand, service operators may be reluctant to require a lot of associated responsibility off the motive force once there's still a choice to interfere so legal problems are seen as a significant challenge within the adoption section, whereas the case may become easier once a substantial share of AVs is reached.

Smith (2012, 2013) examines the legal implications of AVs in the United States, concluding that they are likely legal as long as a driver can maintain control of the vehicle at all times. While considering into viable legal and liability eventualities, together with privacy problems (Schoonmaker, 2016), as mentioned in law officer (2016), Schellekens (2015), Lederman et al. (2016). China, South Korea, the United States, and therefore the EU have displayed no response in addressing such issues related to liability and insurance risks of AVs. The US national has not created any rules but urged the regime to require action in allocation liability and vehicle insurance (NHTSA, 2017). The EU remains exploring Jewish calendar month liability risks, and therefore, the European parliament created recommendations in 2017 to form a required insurance theme for victims of Jewish calendar month accidents and creating robots wrongfully in control of accidents (EP, 2017).

6.2.4 VALUES

It should also be remembered that in situations where a crash is unavoidable, the decisions made by AVs may be questioned in the court. A human motorist is not held accountable if he or she must choose between hitting a buck crossing the road, colliding with a car approaching from the reverse direction, or crashing into a forest. When faced with the choice of hitting a buck crossing the roadway, hitting with a car traveling in the reverse direction, or slamming into a forest, human drivers are frequently not held responsible for the split-second decision they must make.

Driving, according to Goodall (2014), necessitates continuous risk evaluations and, as a result, the necessity to make ethically and legally difficult decisions. He recommends further moral study in AV decision-making systems and dismisses nine responses. In an enhancement work that can be updated in AVs, Gerdes and Thornton (2015) present behaviors in which moral contemplations can be converted into scientific cost or limitations. Finally, when designing AVs, there is a need for more transparency and consideration in making these moral decisions. They shouldn't be taken lightly, and they shouldn't be made solely by manufacturers. Alternatively, perhaps the general public should be involved or, at the very least, made aware of the moral decisions that AVs are updated to make. Furthermore, not giving enough care to this issue while presenting AVs on public roads could pose a serious difficulty for independent vehicles in the event of a catastrophic disaster, whether owing to manufacturers' negligence or any consequent open anti-innovation sentiment.

6.2.5 HUMAN ENGINEERING

As previously stated, fully autonomous Level 4 vehicles capable of driving themselves anywhere will most likely take some time to become commercially available. Vehicles that can drive themselves at a specific distance, under specific conditions, and on particular streets may be launched in the next years. In some circumstances, a human driver will be required to assist these cars. Because the automation system and the human driver will both be in charge of the vehicle, it's crucial to anticipate how the human administrator and the AV will communicate. This is something that many human components involved in AV consider.

6.3 AUTONOMOUS VEHICLE IN LOGISTICS: UTILIZATION AND OUTCOME

The final section offered a wide perspective and insight into the wider context of autonomous vehicle adoption, allowing researchers to better understand the future applications, impacts, and challenges of autonomous vehicle adoption in logistics. The value of AVs for supply is the subject of this segment. As previously stated,

Use of Autonomous Vehicles in Logistics

logistics is described in this paper as the structures put in place by organizations to transfer merchandise consistently between geographical locations. As previously stated, the work on AVs focuses on technological aspects, possible roadblocks, advantages, and costs. Furthermore, the emphasis should be primarily toward autonomous passenger vehicles, ignoring a substantial portion of road transport that involves the transportation of various types of goods (Flämig, 2015). As a result, the introduction of AVs in logistics, and thus their implications, has received relatively little publicity thus far. Despite the lack of study, there are a variety of reasons why AVs may be used sooner in logistics than in passenger transportation (Ghaffary, 2014; Stromberg, 2014). First and foremost, as DHL (2014) points out, the greatest setting for AVs is usually supply. It is considerably easier to run AVs in controlled environments like factories, processing plants, or harbors, as well as in remote out-of-door locations, than in the advanced setting of metropolitan traffic. Furthermore, in these circumstances, the employment of AVs is subject to fewer laws and restrictions.

Since logistics practitioners are familiar with those environments and can benefit from using AVs, they can complete the generation faster in regular site visitors than customers who are unfamiliar with the generation. Second, it has long been argued that when transporting objects rather than individuals, the aforementioned liability problem may be less serious. Finally, companies are more likely to make decisions based on future cost savings, while consumers may be more open to issues of faith and ethics. Furthermore, as will become apparent in this section, self-sustaining motors are expected to have a significant impact on logistics as we know it today. With the adoption of AVs, across the whole supply chain, logistical operations, from raw material mining to intermediate delivery, maneuvers in warehouses, distribution centers, and manufacturing plant life all the way to technologies that connect the final mile, should be laid low, freeing up room for entirely new business models.

6.3.1 AVs in Indoor Logistics: Applications

The term "autonomous vehicle" has primarily been applied for vehicles with an automation level of three or four, as well defined by the National Highway Traffic Safety Administration (2014. However, indoor logistics described in this section, the term "automated guided vehicle" frequently used in the many literature (AGV). AGVs are designated as "autonomous vehicles that are widely used to transport materials between workstations in a variety of manufacturing systems and perform a variety of tasks involving automation in industrial environments" (Kalinovcic et al., 2011; Vivaldini et al., 2015). AVs must be interpreted in this context to mean any vehicle that does not need the assistance of a human driver to operate. With the introduction of AVs, room for new business models will be created, while current ones will be exhausted.

As previously mentioned, AGVs have been used in indoor supply settings for a long time. Barrett Electronics Corporation introduced the first automatic guideway vehicle (AGV) to the market in 1954, which was also the first time car automation was used (Lagorio-Chafkin, 2014; Scribner, 2014). While dragging a trailer comparable to a tow motor, the vehicle, which was originally deployed in a supermarket warehouse, crashed with an overhead wire. Wire technology, in which the vehicle follows radio waves carried through the ground by a wire, is also used by some AGVs.

Victimization guide tape technology is used to produce AGVs, in which a colored, reflective tape is put into the infrastructure to direct vehicles equipped with cameras, sensors, or magnets to the tape (RoboteQ, 2015). Instead of using tape to direct AGVs, alternate visual elements would be built into the infrastructure. The downside of AGV technologies that rely on infrastructure to direct vehicles is that they are limited in their versatility, as these AGVs can only be used in predetermined ways. Furthermore, some of those AGVs don't seem to be capable of traversing a sudden obstacle in their way, resulting in a blockage before a person's operator takes command or an obstruction is removed (Vivaldini et al., 2015). The market for AGVs using vision steerage technology has recently been relaunched. These AGVs use depth cameras, lasers, and sensors to constantly track their surroundings, resulting in a 3D map that can span many preloaded infrastructure components (Möller et al., 2012). This current generation of AGVs is gradually becoming totally autonomous, capable of traveling freely on any available path within an indoor setting, opening up more possibilities than ever before. At the same time, AGV's acceptability in the sector looks to be peaking, as significant manufacturers and distributors have increased their utilization (Lagorio-Chafkin, 2014).

When compared to manual operations, AGVs can increase performance, productivity, and accuracy while also increasing protection (DHL, 2014). Furthermore, as compared to older material handling equipment such as conveyors, carrousels, and ASRS, AGVs have greater flexibility in handling differences in scale, form, product dimensions, weight, and volume, as well as mechanical qualities. They allow to respond more quickly to changes that necessitate a new site layout, avoiding the need for time-consuming retrofitting. Scalability is encouraged by AGVs in order to adjust to growth and deal with seasonal demand. Furthermore, the device remains operational in the case of a malfunction or the need for technical repair of one or more AGVs due to the modularity of the AGVs.

6.3.2 AVS IN OUTDOOR LOGISTICS: APPLICATIONS

Because there is significantly less complexity, fewer regulations apply, liability is not as complex, and corporate rationale driven by efficiency is used, these private outdoor areas, including indoor settings, are best suited to the usage of AVs than public roadways. As a result, as this chapter will demonstrate, AVs of various types are already being employed to perform material handling jobs in ports and logistic yards. AVs are not yet commonly employed at airports, according to DHL (2014), but cargo transporters may be more efficient if vehicle automation technology is implemented.

In any event, it is evident that material handling efficiency is vital to port, airport, and logistics yard productivity. For one thing, revenue is generated only while ships, planes, and trucks are in motion. As a result, production times need to be kept as short as possible. Second, because products cannot be used while in transit, delivery must be finished as quickly as feasible. Given the importance of these two criteria for the competitiveness of airports, ports, and logistics yards, it's not surprising to see a trend of automation comparable to that observed in indoor logistics emerge (Vis, 2006).

As previously stated, airports appear to be lagging behind, but when it comes to transport equipment at harbors, automation is the most essential trend in progress (Carlo et al., 2014). After all, the advantages are the same whether you're inside or outside. AVs can also reduce vehicle wear, fuel consumption, and labor costs by eliminating human error, being incredibly reliable and exact when driving, allowing goods and vehicles to be monitored in real time, and reducing vehicle wear, fuel consumption, and labor expenses (Demuth, 2012).

AGVs currently operate at lower speeds and on fixed paths in secure outside locations, guided by infrastructure and on-board cameras and lasers. Using GPS navigation, for example, improves mobility at the expense of more challenging traffic management, which must avoid deadlocks, crashes, and congestion (Carlo et al., 2014). Container Terminal Altenwerder (CTA) used AGV at Hamburg harbor with a minor amount of effort (OELCHECK GmbH, 2009). They're employed on desolate public territory as a transition between using AVs in protected outside environments and in regular traffic, and they're expanding outside private outdoor territory. As a result, it's no surprise that autonomous vehicles/trucks are now commonplace in the mining business. Rio Tinto employs 53 autonomous trucks across numerous iron ore mines, demonstrating that AVs reduce costs while improving productivity, health, and environmental protection (Coyne, 2015).

6.3.3 Autonomous Vehicles in Long-Haul Freight Transport

In the previous sections, the logistics use cases showcasing current improvements in vehicle automation technology, i.e., in secure private indoor and outdoor areas, were mostly extensions of existing applications. For quite some time, AGVs have been employed for material handling in private and abandoned regions. Over the years, several autonomous truck initiatives have been created, with a focus on platooning. The early European projects, such as PROMOTE CHAUFFEUR I & II (IST World, 2000), concentrated on the required technology, while the University of Aachen's KONVOI project explored the impact of platooning as well as its legal and economic ramifications.

Why should AVs be utilized on public highways for long-haul haulage before being utilized for passenger transportation or on smaller routes? First and foremost, the advantages of not hiring a driver are particularly considerable in the road freight industry, as they are in the driver salary and benefits account in the United States. More than a third of the entire shipping cost is spent on this (Fender & Pierce, 2012). Furthermore, because the highway environment is far more predictable and less dynamic than, say, city streets, autonomous car technology would very definitely be ready for highway driving initially.

6.3.4 Autonomous Vehicles in Highway Trucking

Commercially available fully autonomous vehicles, like AV technology in general, are still a way off. Mercedes-Benz has pledged to commercialize an autonomous vehicle by 2025 as part of its Future Truck 2025 program (Anon, 2014). Other producers made a variety of commitments as well. Many self-driving assistance technologies are already in use in today's automobiles. Consider gadgets that warn the driver about safe driving distances, activity in the vehicle's blind zones, emergency braking, and lane maintaining. DHL (2014) offers several assisted highway trucking scenarios based on these automation capabilities that may impact road freight operators' operations. In the first case, a supported highway trucking system, according to DHL (2014), can autonomously move a truck safely inside its lane.

6.3.5 Contemplation on Logistics Operation

Despite the confusion surrounding the feasibility of integrating AVs into everyday traffic, the majority of industry experts and references polled appear to decide that the question is not whether AVs would be eligible for widespread usage but when. It seems unlikely that the production of AVs will come to a standstill, with practically all major automobile manufacturers and other notable corporations such as Google operating on their AVs and governments investing in research initiatives. Vehicle automation is already evident in many parts of today's cars.

Given the significant impact that AV adoption is supposed to have on logistics best practices, the lack of logistics-related research on AVs that goes beyond dispatch, scheduling, and routing of AVs in secure indoor and outdoor contexts is alarming. The most recent research on the use of automated trucks in long-haul trucking focuses on platooning. The focus is on companies rather than how platoons should be organized or what impact platooning would have on platoon training and route planning. Investigators have a lot of room to improve these models and run simulations and field experiments. There are also questions about the effect of these last-mile-bridging models on businesses' inclusive supply chains and operations.

6.4 AUTONOMOUS VEHICLES IN LOGISTICS: RISKS

To bridge the gap between theory and practice, current challenges from both theory and practice must be identified and implemented. A comprehensive view on threats can be applied in this way. The goal is to categories risks such that a framework may be created that incorporates risk groupings. As a result, dangers may be grouped and detected in these classes in a methodical manner. Furthermore, the framework may be updated if new threats emerge that do not fit into any of the existing categories. As a result, the correlation between a risk category and independent collaboration characteristics can indeed be stated and shown. This study follows Harland et al.'s (2014) risk classification, which summarizes many characteristics to present a comprehensive and multiperspective view of hazards.

The various dangers associated with the use of AVs in current logistics system is being briefly discussed as under.

Operation risks	According to Meulbroek (2016), they are incidents that influence an organization's internal capacity to manufacture things and offer services at the aggregate level (e.g., machine breakdowns or employee absenteeism). According to Simons (2014), operational dangers are "the consequences of a failure in the essential competency of service, manufacture, or processing"

(Continued)

Cybersecurity	In 2013, the EU adopted the cybersecurity policy, and in 2016, the NIS Directive was enacted to handle cybersecurity (European Commission, 2017). The EU Agency for Network and Data Protection (ENISA, 2016) also issued guidelines on how to improve cybersecurity in non-AVs that are linked. The North American country SPY Car Act (2017) improves vehicle cybersecurity by enforcing device standards, such as securing data at all stages of storage and transmission and requiring all vehicles to have malfunction detection capabilities
Supply and customer risks	Input risks, often known as supply risks, affect inbound flows of any form of resource necessary for operations (e.g., absence of delivery of raw materials). Consumer hazards include things like product obsolescence, which might affect the likelihood of customer orders (e.g., misunderstanding if an order is placed by customers)
Competitive risks and reputation risks	Competitive hazards develop when a company's capacity to differentiate its products/services from those of rivals is limited (e.g., the more generic the products/services are, the more difficult it is to distinguish). According to Schwartz and Gibb, reputation risks are a reduction in a company's valuation due to a loss of customer confidence (e.g., environmental pollution caused by a firm)
Financial risk and fiscal risks	Financial hazards put a corporation at danger of losing money due to market swings (e.g., changes in currency rates). They can also happen as a result of a debtor's failure to pay (e.g., debtor insolvency). Tax changes that have an impact on a company's financial condition might lead to fiscal hazards (e.g., enhancement of sales tax or release of new taxes)
Regulatory risk and legal risks	Regulatory risks impacting a company's operations due to changes in legislation such as environmental protection are the responsibility of Bowen et al. (2014), Smallman (2012), and Meulbrook (2016) (e.g., limitation of CO_2-emission). As a result of potential disagreements with customers, suppliers, shareholders, or employees, legal risks expose a company to lawsuit (e.g., through insufficient challenging of consumer intemperance)

6.5 CONCLUSION

This paper gives researchers and logistics professionals a broad introduction to AVs (cars/trucks), including an assessment of the state-of-the-art and its implications for the logistics industry. The effect of autonomous collaboration technology implementation and use on potential risks has been addressed. The use of autonomous cooperation to reduce logistics risks has some potential positive effects, but it also has the potential to increase risks. The first section of the paper addressed history, emerging technologies, liabilities, ethics, and legislation, while the second session focused on the benefits and outcomes of AVs in recent logistic systems, as well as the risks that come with them. In conclusion, outcomes can be influenced (positive: opportunities and negative: risks); however, net effects are simply debated rather than measured.

REFERENCES

Anderson, J. et al., 2014. *Autonomous Vehicle Technology: A Guide for Policymakers.* Santa Monica, CA: Rand Corporation.
Carlo, H. J., Vis, I. F. A. & Roodbergen, K. J., 2014. Transport operations in container terminals: Literature overview, trends, research directions and classification scheme. *European Journal of Operational Research*, 236(1), pp. 1–13.

Casey, M., 2014. Want a self-driving car? Look on the driveway [Online]. Available at: http://fortune.com/2014/12/06/autonomous-vehicle-revolution/ [Accessed 22 June 2015].

Dagan, E., Mano, O. & Stein, G. P. S. A., 2004. Forward collision warning with a single camera. *IEEE Intelligent Vehicles Symposium, 2004*, Parma, Italy.

Demuth, R., 2012. Fahrerloser LKW in einer Molkerei [Online]. Available at: http://www.goetting.de/news/2012/molkerei [Accessed 7 July 2015].

DHL, 2014. Self-driving vehicles in logistics: A DHL perspective on implications and use cases for the logistics industry. s.l:s.n.

Fender, K. J. & Pierce, D. A., 2012. *An Analysis of the Operational Costs of Trucking: A 2012 Update*. Arlington: American Transportation Research Institute.

Flämig, H., 2015. Autonome Fahrzeuge und autonomes Fahren im Bereich des Gütertransportes. In: *Autonomes Fahren: Technische, rechtlige und gesellschaftliche Aspekte*. Cham, s.l: Springer.

Ghaffary, S., 2014. Robot roundup: How supply chain is leading the way in sophisticated automation [Online]. Available at: http://blog.elementum.com/robot-roundup-how-supply-chain-is-leading-the-wayin-sophisticated-automation [Accessed 25 June 2015].

Goodall, N. J., 2014. Machine ethics and automated vehicles. In: Meyer, G. & Beiker, S. (eds) *Road Vehicle Automation*. Cham: Springer, pp. 93–102.

Kalinovcic, L., Petrovic, T., Bogdan, S. & Bobanac, V., 2011. Modified Banker's algorithm for scheduling in multi-AGV systems. *2011 IEEE International Conference on Automation Science and Engineering*, Trieste, Italy, pp. 351–356.

Lagorio-Chafkin, C., 2014. Automated guided vehicles: Behind the swift business of a heavy industry [Online]. Available at: http://www.inc.com/christine-lagorio/best-industries/automated-guided-vehicles.html [Accessed 29 June 2015].

Möller, A. et al., 2012. A mobile indoor navigation system interface adapted to vision-based localization. Proceedings of the 11th International Conference on Mobile and Ubiquitous Multimedia, ACM, New York.

National Highway Traffic Safety Administration, 2013. *Preliminary Statement of Policy Concerning Automated Vehicles*. Washington, DC: National Highway Traffic Safety Administration.

RoboteQ, 2015. Building a magnetic track guided AGV. [Online] Available at:http://www.roboteq.com/index.php/applications/applications-blog/building-a-magnetic-track-guidedagv[Accessed 29 June 2015].

Scribner, M., 2014. Human achievement of the day: Autonomous vehicles, from imagination to reality [Online]. Available at: https://cei.org/blog/human-achievement-day-autonomous-vehicles-imagination-reality [Accessed 29 June 2015].

Smith, B. W., 2013. *Automated Driving: Legislative and Regulatory Action*. Stanford, CA: The Center for Internet and Society.

Stromberg, J., 2014. Why trucks will drive themselves before cars do [Online]. Available at: http://www.supplychain247.com/article/why_trucks_will_drive_themselves_before_cars_do [Accessed 22 June 2015].

Vahidi, A. & Eskandarian, A., 2003. Research advances in intelligent collision avoidance and adaptive cruise control. *IEEE Transactions on Intelligent Transportation*, 4(3), pp. 143–153.

7 Assessment of Challenges for Implementation of Industrial Internet of Things in Industry 4.0

Snigdha Malhotra, Tilottama Singh, and Vernika Agarwal
Amity University

CONTENTS

7.1 Introduction 127
7.2 Review of Existing Literature 128
7.3 Research Methodology 129
7.4 Numerical Illustration 133
    7.4.1 Research Design 133
    7.4.2 Application of the Grey DEMATEL Approach 134
    7.4.3 Result and Analysis 138
7.5 Conclusion and Future Scope 138
References 138

7.1 INTRODUCTION

Information technology (IT) has grown at a tremendous pace in the recent years globally. The Indian industries are also matching the pace of this development in the arena of Industry 4.0 which bases itself on Industrial Internet of Things (IIoT) which is an efficient system for enhancing the effective utilization of cyber system, smart devices, and IT with the aim of making intelligent decisions (Agrawal et al., 2019). With the transformation of economies to circular economy aiming to base itself on sustainability and reuse of existing resources, it has created a path toward IIoT, cloud computing, and more of human-machine interaction for effective working.

Although a number of researchers have worked in the area of Industry 4.0, there is a substantial gap in examining the challenges of implementation of IIoT techniques, especially in context of India and also reflecting on their cause-effect relation for successful future implementation. The revolution in the IIoT has made plants, products, and systems more connected (Boyes et al., 2018). An unexceptional combination of IT and operational technology (OT) in industrial environments forms the basis for transparent and efficient processes working on a higher degree of atomization and precision (Thames and Schaefer, 2016). There has been a lot of groundwork undertaken by the Indian economy in the form of policies and plans supporting the implementation of IIoT which is the future of business, yet there is a substantial gap when we examine the implantation and result of IIoT techniques in Indian industries, as rightly studies have demonstrated that prior changes, planning and reframe of the system are essential to ensure any changes, aiming to reduce the failure rate (Wortmann and Flüchter, 2015). The research promises a lot of significance as India has a huge capability for the development of its IIoT industry.

It can be seen from the above discussion that there is a need to understand the challenges in implementing IIoT techniques in the Indian industry and to understand the interrelationship among them. The paper therefore examines the following research question:

- What are the major difficulties or problems in successful implementation of IIoT?
- What is the interrelationship between various challenges?

The present chapter aims to understand the challenges faced by Indian enterprises in India in implementing IIoT in their business operations. The methodology of grey decision-making trial and evaluation laboratory (DEMATEL) is employed to understand the interrelationship between these challenges and to identify them as cause and effect. The rest of the chapter is divided into the following sections: Section 7.2 discusses the existing literature followed by research methodology in Section 7.3. The numerical discussion is elaborated in Section 7.4 followed by conclusion and future scope in Section 7.5.

7.2 REVIEW OF EXISTING LITERATURE

The IIoT which is also referred to as Industry 4.0 aims at integrating technologies into industrial value chain, operations, and work, connecting the world digitally (Hartmann and Halecker, 2015), and its adoption has drastically filled the gaps in contemporary challenges of the global economy. In simple words, we can say that it is effective in preparing the companies for future work scenarios and is required in present and future (Arnold et al., 2016). The future of the sustainable and successful organization lies drastically on the successful adoption and implementation of IIoT in organizations (Sengupta et al., 2020). Innovation and transformations go hand in hand, but successful firms can only happen when they are prepared to cope with resistance to change and adapt to the innovative techniques effectively (Pisano, 2015). At times, these challenges can also be a part of the culture of such organizations, and therefore, new technologies related to IIoT though carry the advantage of better output and efficiency but also require new competencies and core development for providing a necessary platform for successful implementation, which is an essential requirement prior to execution (Delery and Doty, 1996). There is scant research in the area of IIOT, specifically identifying challenges in implementation of IIoT. IIOT brings new sustainability prospects, increased efficiency, more flexibility but also requires trained workforce, data security, expanding business, and may be other factors with specific industries (Kagermann et al., 2011). Seeing the economic perspective, the IIoT facilities enhance the connectivity of business, reduce costs, and create better quality of service and products (Oesterreich and Teuteberg, 2016), and all these together lead to increased customer satisfaction (Stock and Seliger, 2016). However, like any other technological implementation, even successful implementation of IIoT requires certain precondition generations like the foremost need of very huge capital and even an economy, which supports such heavy capital investment (Laudien and Daxbock, 2016). Also a strong transparent network and appropriate logistics can reduce cost to certain level (Zhou et al., 2017), and for all this, the organizations need to be really proactive and prepared (Tesch et al, 2017) which is the grey area or the research gap, taken in the current study.

Though the empirical studies have emphasized on importance, challenges, and discussions on advantages, the literature lacks in developing comprehensive economic analysis on the challenges in successful implementation of these IIoT techniques in different manufacturing companies. The authors aim in overcoming these gaps by analyzing the prechallenges of IIoT, disclosing their perception with reference to specific industry. Scant research has led us to adopt grey DEMETAL design and explain the most dominating challenges through this approach.

7.3 RESEARCH METHODOLOGY

Business applications extensively use the DEMATEL method. The cause-effect relationships among the factors are effectively revealed through classical or crisp DEMATEL. It also helps in prioritizing the factors. It may have certain challenges in describing uncertainties. The extensions of the DEMATEL method are used in order to overcome this challenge and to increase its capabilities.

In real-world applications, the ambiguity increases due to unsuitable human judgements and vague information. These inaccurate sources include unquantifiable, incomplete, and difficult-to-obtain information and partial ignorance. Since these limitations cannot be effectively overcome by classical or crisp DEMATEL, the grey system theory-based DEMATEL method would be an appropriate and effective approach. Deng-developed grey system theory is a methodology that helps effective integration of uncertainty and ambiguity into the evaluation process. It is a methodology for efficient analysis of systems with imprecise information and is enabled to handle uncertainty successfully.

Computational steps of the grey-based DEMATEL are given below.

Step 1: Identification of the evaluation criteria and the grey linguistic scale: To represent the uncertainties of human assessments, we determine and recognize the evaluation criteria and a grey linguistic scale. The linguistic scale and the corresponding grey numbers are as follows: "No Influence (NI)" is represented as [0, 0], "Very Low Influence (VL)" as [0, 0.25], "Low

Influence (L)" as [0.25, 0.5], "High Influence (H)" as [0.5, 0.75], and "Very High Influence (VH)" as [0.75, 1].

Step 2: Determining the direct-relation matrix: To quantify the relationship between criteria shown as $C = \{C_i | i = 1, 2,\ldots, n\}$, a group of experts are requested to make pair-wise comparisons in terms of linguistic scale given in Table 7.1. The initial direct-relation grey matrix A is obtained.

TABLE 7 1
List of Shortlisted Challenges

Notation	Challenges	Description	References
C1	Lack of efficient energy resources	Many IIoT applications need to run for years on batteries. This calls for the design of low-power sensors which do not need battery replacement over their lifetimes. This creates a demand for energy-efficient designs	Sisinni et al. (2018)
C2	Real-time performance	IIoT devices are typically deployed in noisy environments for supporting mission- and safety-critical applications, and have stringent timing and reliability requirements on timely collection of environmental data	Sisinni et al. (2018)
C3	Coexistence and interoperability	There will be many coexisting devices arranged in close proximity in the limited spectrum with rapid increase in IIoT connectivity. This brings forward the important challenge of coexistence in the crowded ISM bands. So, interference between devices must be taken care of to keep them working	Sisinni et al. (2018)
C4	Security challenge	All devices whether industrial machines, computer, tablets, or smart phones need to be updated on regular basis whether to avoid threats or due to configuration changes installed in these devices spread across the geographical location or inside factory	Zetter (2011)
C5	Privacy challenge	It refers to platform security, secure engineering, security management, identity management, and industrial rights management	Sadeghi et al. (2015)
C6	Chaos challenge	IIoT lacks in developing coordination among the varied departments in organizations, to use the data	Kumar and Iyer (2019)
C7	Data management challenge	Lack of a standardized approach for data management is still one of the concerns in big companies. Specifically, there is a problem in which a plethora of intermediate solutions exist for data management within a company; it ranges from storing and exchanging data in the form of printouts, emails, excel sheets, proprietary applications, and using heterogeneous database solutions between various departments or production halls	Khan and Turowski (2016)

(Continued)

TABLE 7 1 (*Continued*)
List of Shortlisted Challenges

Notation	Challenges	Description	References
C8	Data mining challenge	Lack of proper data management and its utilization proactively is still one of the major challenges. The proper processing of data making the best use for planning and operations remains a grey area in most of the organizations	Kumar and Iyer (2019)
C9	Lack of organization and production fit	Effective connectivity and synchronization still lack in successful integration of IT and OT	Kumar and Iyer (2019)
C10	Lack of future viability during implementation	There still exist a gap between implementation of IIoT and its future viability in most of the organizations. The planning and implementation lack in developing a futuristic approach in proper implementation of IIoT to avoid further cost and make it more sustainable	Munirathinam (2020)
C11	Lack of proactive maintenance	Integration between economy and global requirements	–
C12	Connectivity and visibility	Internet outages, power loss, and manual and technical errors create frequent troubles. This leads to the removal of device from connected digital network, which adversely effects the production process and requires extra cost to overcome the damage caused	Kumar and Iyer (2019)

$$A^k = \begin{matrix} C_1 \\ C_2 \\ \vdots \\ C_n \end{matrix} \begin{bmatrix} 0 & \otimes a_{12}^k & \cdots & \otimes a_{1n}^k \\ \otimes a_{21}^k & 0 & \cdots & \otimes a_{2n}^k \\ \vdots & \vdots & \cdots & \vdots \\ \otimes a_{n1}^k & \otimes a_{n2}^k & \cdots & 0 \end{bmatrix} \quad (7.1)$$

where

k is the number of experts, $=[\underline{a}_{ij}, \overline{a}_{ij}]$ are grey numbers and $= [0, 0]$ for $i = 1, 2,\ldots, n$.

Step 3: Combination of all grey direct-relation matrices: All the grey direct-relation matrices obtained are averaged by using Eq. (7.2), and we get the aggregate matrix Z.

$$Z = \left(\sum_{i=1}^{k} A^k\right) \bigg/ k \quad (7.2)$$

Step 4: Determining the structural model: In this step, the transformation of criteria scales into comparable scales is carried out, and the linear scale transformation is changed to a normalization formula. Let

$$\sum_{j=1}^{n} \otimes z_{ij} = \left[\sum_{j=1}^{n} \underline{z_{ij}}, \sum_{j=1}^{n} \overline{z_{ij}} \right] \quad (7.3)$$

And $r = \max_{1 \leq i \leq n} \left(\sum_{j=1}^{n} \overline{z_{ij}} \right)$.

Then, the normalized direct-relation grey matrix, G, is equal to $G = r^{-1} \times Z$
And

$$G = \begin{bmatrix} 0 & \otimes g_{12} & \cdots & \otimes g_{1n} \\ \otimes g_{21} & 0 & \cdots & \otimes g_{2n} \\ \vdots & \vdots & \cdots & \vdots \\ \otimes g_{n1} & \otimes g_{n2} & \cdots & 0 \end{bmatrix}$$

Where $\otimes g_{ij} = \dfrac{\otimes z_{ij}}{r} = \left[\dfrac{\underline{z_{ij}}}{r}, \dfrac{\overline{z_{ij}}}{r} \right]$ \quad (7.4)

Step 5: Establishing the total relation matrix: After getting matrix G, the grey normalized direct relation, the relation matrix of grey total T can be found by using the following equations:

$$T = G + G^2 + \ldots + G^k$$
$$T = G(I - G)^{-1}, \quad \text{when } \lim_{k \to \infty} G^k = [0]_{n \times n} \quad (7.5)$$

$$T = \begin{bmatrix} \otimes t_{11} & \otimes t_{12} & \cdots & \otimes t_{1n} \\ \otimes t_{21} & \otimes t_{21} & \cdots & \otimes t_{2n} \\ \vdots & \vdots & \cdots & \vdots \\ \otimes t_{m1} & \otimes t_{m2} & \cdots & \otimes t_{mn} \end{bmatrix}$$

And $\otimes t_{ij} = \left[\underline{t_{ij}}, \overline{t_{ij}} \right]$ \quad (7.6)

And $\text{Matrix}\left[\otimes \underline{t_{ij}} \right] = \underline{G} \times (I - \underline{G})^{-1}$

$\text{Matrix}\left[\otimes \overline{t_{ij}} \right] = \overline{G} \times (I - \overline{G})^{-1}$

Step 6: Whitenization: The grey total matrix T is whitened before calculating the sum of rows and columns. The grey numbers are changed to crisp values by the modified Converting Fuzzy data into Crisp Scores (CFCS) method given below.

Implementation of Industrial IoT in Industry 4.0

$$\otimes \underline{t_{ij}} = \left(\otimes \underline{t_{ij}} - \min \otimes \underline{t_{ij}}\right) \big/ \Delta_{\min}^{\max} \quad (7.7)$$

$$\otimes \overline{t_{ij}} = \left(\otimes \overline{t_{ij}} - \min \otimes \overline{t_{ij}}\right) \big/ \Delta_{\min}^{\max} \quad (7.8)$$

Where $\Delta_{\min}^{\max} = \max \otimes \overline{t_{ij}} - \min \otimes \underline{t_{ij}}$

$$Y_{ij} = \frac{\otimes \underline{t_{ij}} \left(1 - \otimes \underline{t_{ij}}\right) + \otimes \overline{t_{ij}} \times \otimes \overline{t_{ij}}}{1 - \otimes \underline{t_{ij}} + \otimes \overline{t_{ij}}} \quad (7.9)$$

$$z_{ij} = \min \otimes \underline{t_{ij}} + Y_{ij} \Delta_{\min}^{\max} \quad (7.10)$$

where
z_{ij} are the crisp values.

Then the sums of rows and columns are distinctly denoted as d and r within the total relation matrix T as in Eq. (7.11).

$$T = \left[t_{ij}\right], \quad i,j \in \{1,2,\ldots,n\}$$

$$d = (d_i)_{n \times 1} = \left[\sum_{j=1}^{n} t_{ij}\right]_{n \times 1}$$

and

$$r = (r_j)_{1 \times n} = \left[\sum_{j=1}^{n} t_{ij}\right]_{1 \times n} \quad (7.11)$$

Step 7: Analyzing the results: The value of $d + r$ sum shows the effects among criteria, and $d - r$ shows the causal relations among criteria.

7.4 NUMERICAL ILLUSTRATION

The application of the grey DEMATEL approach is illustrated in two phases – research design and data collection. The detail of each phase is discussed in Sections 7.4.1 and 7.4.2.

7.4.1 RESEARCH DESIGN

The objective behind the research design is to recognize the challenges faced by enterprises in India in implementing IIoT in their business operations. An extensive literature survey and interview of various stakeholders were carried out to find the list of challenges. Seven stakeholders were identified for the interpretation of the data.

The team of stakeholder comprises three academicians and four industry professions working in the field of IIoT. This assorted team of stakeholders was selected to incorporate various perspectives into the decision-making process. Based on the consensus of these stakeholder, the following challenges were shortlisted as given in Table 7.1.

The aim of the chapter is to understand the challenges faced by the Indian enterprises in implementing IIoT in their business operations using the grey DEMATEL methodology. To conduct the study, challenges were shortlisted which will lead to the achievement of the objectives.

7.4.2 Application of the Grey DEMATEL Approach

The proposed methodology is used to analyze the challenges for IIoT adoption. The methodology is used as per the steps illustrated above. After finalizing the list of the challenges faced by the companies based on previous research studies and discussions with stakeholders, the data were analyzed. Following the steps of research methodology, the evaluation of the challenges was done, and linguistic scale was assigned to the motivational factors. Table 7.2 represents the evaluation of challenges by the stakeholders.

In the similar way, the data were collected from the rest of the decision makers.

It can be seen in Table 7.3 that all the direct-relation grey matrices are obtained and averaged. The average aggregate matrix Z is derived. To convert criteria scales to a comparable scale, the linear scale is transformed into a normalized formula using step 4. In Table 7.4, the normalized direct-relation matrix G is illustrated.

After we obtain the normalized direct-relation grey matrix, now using step 5, we obtain the grey total relation matrix T.

After the grey total relation matrix T is obtained, the matrix T is whitened to get the crisp values using the CFCS method. The whitening of matrix T is done using the step 6 of methodology. After the values are converted to crisp values, the

TABLE 7.2
Linguistic Term Matrix for First Decision Maker

	C1	C2	C3	C4	C5	C6	C7	C8	C9	C10	C11	C12
C1	No	No	H	VL	VL	L	L	H	No	H	VH	VH
C2	VL	No	No	H	VH	VH	VL	H	VL	VH	H	H
C3	H	VL	No	VL	VL	VL	H	H	VL	VL	H	L
C4	VL	H	L	No	VH	H	VL	H	L	H	L	H
C5	VL	H	H	H	No	H	VL	L	VL	H	VH	H
C6	L	H	L	H	H	No	VL	H	L	H	H	VL
C7	H	L	H	VL	VL	L	No	H	H	L	L	L
C8	H	H	H	H	H	H	L	No	L	L	L	L
C9	VL	H	VL	VL	VL	H	L	H	No	H	H	H
C10	L	H	VL	H	VH	H	VL	H	L	No	H	L
C11	VH	H	VH	L	H	H	H	L	VH	VH	NO	VH
C12	VH	VH	H	VH	VH	H	H	L	VH	VH	VH	No

TABLE 7.3
Average Aggregate Matrix Z

	C1	C2	C3	C4	C5	C6	C7	C8	C9	C10	C11	C12
C1	[0, 0]	[0.125]	[0.625, 0.875]	[0.063, 0.313]	[0.375, 0.563]	[0.188, 0.375]	[0.438, 0.688]	[0.5, 0.75]	[0, 0]	[0.25, 0.5]	[0.75, 1]	[0.688, 0.938]
C2	[0.188, 0.438]	[0, 0]	[0, 0.063]	[0.5, 0.75]	[0.313, 0.563]	[0.25, 0.5]	[0.375, 0.625]	[0.438, 0.688]	[0.125, 0.313]	[0.438, 0.625]	[0.375, 0.625]	[0.313, 0.563]
C3	[0.563, 0.813]	[0, 0.188]	[0, 0]	[0, 0.25]	[0, 0.188]	[0.063, 0.25]	[0.313, 0.563]	[0.375, 0.625]	[0, 0.188]	[0, 0.188]	[0.5, 0.75]	[0.188, 0.375]
C4	[0.125, 0.375]	[0.438, 0.625]	[0.063, 0.313]	[0, 0]	[0.563, 0.813]	[0.125, 0.313]	[0.313, 0.563]	[0.25, 0.5]	[0.313, 0.5]	[0.25, 0.5]	[0.375, 0.625]	[0.563, 0.813]
C5	[0.125, 0.375]	[0.438, 0.688]	[0.25, 0.5]	[0.125, 0.375]	[0, 0]	[0.25, 0.5]	[0.313, 0.563]	[0.063, 0.313]	[0.125, 0.375]	[0.375, 0.625]	[0.625, 0.875]	[0.563, 0.813]
C6	[0.375, 0.625]	[0.375, 0.625]	[0.375, 0.625]	[0.125, 0.375]	[0.438, 0.688]	[0, 0]	[0.438, 0.688]	[0.313, 0.563]	[0.375, 0.625]	[0.438, 0.625]	[0.5, 0.75]	[0.063, 0.313]
C7	[0.250, 0.5]	[0.313, 0.563]	[0.438, 0.688]	[0.125, 0.313]	[0.125, 0.313]	[0.063, 0.25]	[0, 0]	[0.25, 0.5]	[0.375, 0.563]	[0.188, 0.438]	[0.375, 0.625]	[0.125, 0.313]
C8	[0.5, 0.75]	[0.25, 0.5]	[0.313, 0.563]	[0.375, 0.563]	[0.375, 0.563]	[0.125, 0.313]	[0.188, 0.438]	[0, 0]	[0.188, 0.375]	[0.313, 0.5]	[0.188, 0.375]	[0.188, 0.375]
C9	[0.063, 0.25]	[0.375, 0.625]	[0, 0.188]	[0, 0.188]	[0, 0.188]	[0.375, 0.625]	[0.188, 0.438]	[0.375, 0.625]	[0, 0]	[0.375, 0.625]	[0.5, 0.75]	[0.5, 0.75]
C10	[0.125, 0.375]	[0.375, 0.625]	[0.25, 0.5]	[0.125, 0.375]	[0.188, 0.438]	[0.5, 0.75]	[0.375, 0.625]	[0.375, 0.625]	[0.313, 0.5]	[0, 0]	[0.5, 0.75]	[0.25, 0.5]
C11	[0.438, 0.688]	[0.563, 0.813]	[0.688, 0.938]	[0.375, 0.625]	[0.375, 0.625]	[0.375, 0.625]	[0.5, 0.75]	[0.25, 0.5]	[0.375, 0.625]	[0.75, 1]	[0, 0]	[0.625, 0.875]
C12	[0.438, 0.688]	[0.688, 0.938]	[0.563, 0.813]	[0.625, 0.875]	[0.750, 1]	[0.375, 0.625]	[0.375, 0.625]	[0.375, 0.625]	[0.75, 1]	[0.375, 0.625]	[0.625, 0.875]	[0, 0]

TABLE 7.4
Normalized Direct-Relation Grey Matrix G

	C1	C2	C3	C4	C5	C6	C7	C8	C9	C10	C11	C12	C12
C1	[0, 0]	[0, 0]	[0.014, 0.029]	[0.072, 0.101]	[0.007, 0.036]	[0.043, 0.065]	[0.022, 0.043]	[0.050, 0.079]	[0.058, 0.086]	[0, 0]	[0.029, 0.058]	[0.086, 0.115]	[0.079, 0.108]
C2	[0.022, 0.05]		[0, 0]	[0, 0.007]	[0.058, 0.086]	[0.036, 0.065]	[0.029, 0.058]	[0.043, 0.072]	[0.050, 0.079]	[0.014, 0.036]	[0.050, 0.072]	[0.043, 0.072]	[0.036, 0.065]
C3	[0.065, 0.094]		[0, 0.022]	[0, 0]	[0, 0.029]	[0, 0.022]	[0.007, 0.029]	[0.036, 0.065]	[0.043, 0.072]	[0, 0.022]	[0, 0.022]	[0.058, 0.086]	[0.022, 0.043]
C4	[0.014, 0.043]		[0.05, 0.072]	[0.007, 0.036]	[0, 0]	[0.065, 0.094]	[0.014, 0.036]	[0.036, 0.065]	[0.029, 0.058]	[0.036, 0.058]	[0.029, 0.058]	[0.043, 0.072]	[0.065, 0.094]
C5	[0.014, 0.043]		[0.05, 0.079]	[0.029, 0.058]	[0.014, 0.043]	[0, 0]	[0.029, 0.058]	[0.036, 0.065]	[0.007, 0.036]	[0.014, 0.043]	[0.043, 0.072]	[0.072, 0.11]	[0.065, 0.094]
C6	[0.043, 0.072]		[0.043, 0.072]	[0.043, 0.072]	[0.014, 0.043]	[0.050, 0.079]	[0, 0]	[0.050, 0.079]	[0.036, 0.065]	[0.043, 0.072]	[0.050, 0.072]	[0.058, 0.086]	[0.007, 0.036]
C7	[0.029, 0.058]		[0.036, 0.065]	[0.050, 0.079]	[0.014, 0.036]	[0.014, 0.036]	[0.007, 0.029]	[0, 0]	[0.029, 0.058]	[0.043, 0.065]	[0.022, 0.5]	[0.043, 0.072]	[0.014, 0.036]
C8	[0.058, 0.086]		[0.029, 0.058]	[0.036, 0.065]	[0.043, 0.065]	[0.043, 0.065]	[0.014, 0.036]	[0.022, 0.05]	[0, 0]	[0.022, 0.043]	[0.036, 0.058]	[0.022, 0.043]	[0.022, 0.043]
C9	[0.007, 0.029]		[0.043, 0.072]	[0, 0.022]	[0, 0.022]	[0, 0.022]	[0.043, 0.072]	[0.022, 0.05]	[0.043, 0.072]	[0, 0]	[0.043, 0.072]	[0.058, 0.086]	[0.058, 0.086]
C10	[0.014, 0.043]		[0.043, 0.072]	[0.029, 0.058]	[0.014, 0.043]	[0.022, 0.050]	[0.058, 0.086]	[0.007, 0.036]	[0.043, 0.072]	[0.036, 0.058]	[0, 0]	[0.058, 0.086]	[0.029, 0.058]
C11	[0.050, 0.079]		[0.065, 0.094]	[0.079, 0.108]	[0.043, 0.072]	[0.043, 0.072]	[0.043, 0.072]	[0.058, 0.086]	[0.029, 0.058]	[0.043, 0.072]	[0.086, 0.115]	[0, 0]	[0.072, 0.1]
C12	[0.050, 0.079]		[0.079, 0.108]	[0.065, 0.094]	[0.072, 0.101]	[0.086, 0.115]	[0.043, 0.072]	[0.043, 0.072]	[0.043, 0.072]	[0.086, 0.1]	[0.043, 0.072]	[0.072, 0.1]	[0, 0]

cause-effect relationship is obtained between the motivational factors by $(d+r)$ and $(d-r)$ in Table 7.5 given below. The values of $d-r>0$ represent that the motivational factor is a cause, and the values of $d-r<0$ represent that the motivational factor is an effect.

Figure 7.1 represents the relationship between the challenges based on the linguistic scale values given by the key decision maker. The values above 0 represent the causes, and the values below 0 indicate the effect due to the causes.

TABLE 7.5
Values of $d+r$ and $d-r$ for Motivational Factors

	D	r	d+r	d−r	Cause/Effect
C1	18.928	17.337	36.265	1.591	Cause
C2	17.062	17.944	35.006	−0.883	Effect
C3	13.240	17.649	30.890	−4.409	Effect
C4	17.446	16.430	33.876	1.017	Cause
C5	17.393	17.403	34.796	−0.011	Effect
C6	18.145	16.509	34.653	1.636	Cause
C7	14.732	17.630	32.362	−2.899	Effect
C8	15.751	17.648	33.400	−1.897	Effect
C9	15.819	16.539	32.358	−0.720	Effect
C10	16.542	17.783	34.325	−1.242	Effect
C11	21.917	19.310	41.227	2.607	Cause
C12	23.291	18.083	41.374	5.208	Cause

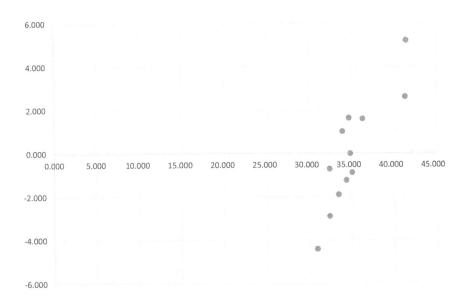

FIGURE 7.1 IRD diagram for challenges.

7.4.3 Result and Analysis

Figure 7.1 represents the relationship of each challenge in implementation of IIoT. The relationship is established by taking out the threshold value and then comparing the values which are greater than the threshold value in each motivational factor listed. As shown in figure 7.1, we can state that C1, C4, C6, C11, and C12 are cause challenges, while C2, C3, C5, C7, C8, and C9 are effects. This helps us understand the relationship between the given factors. C1 – Lack of efficient energy resources, C4 – security challenge, C6 – chaos challenge, C11 – lack of proactive maintenance, and C12 – connectivity and visibility belong to the cause group, which should be controlled and paid more attention to. C2 – real-time performance, C3 – coexistence and interoperability, C5 – privacy challenge, C7 – data management challenge, C8 – data mining challenge, and C9 – lack of organization and production fit are in the effect group that needs to be improved.

7.5 CONCLUSION AND FUTURE SCOPE

The study provides an analytical framework to the stakeholders and decision makers in the IIoT sector. The research focused on scrutinizing the challenges with the aid of various stakeholders. IIoT or Industry 4.0 can help in bringing together brilliant machines, advanced analytics, and the people involved at work. The foremost issue in planning IIoT implementation is understanding the challenges faced by the decision makers in its application. The motivation behind the present study is to equip the Indian industries to understand the benefits of IIoT and overcome the hurdles in its implementation.

The study adopts methodology based on grey DEMATEL to understand the cause-effect relationship among the challenges and to avoid the uncertainty and ambiguity of data. The aim of the paper was to identify the challenges and to establish the relationship for the challenges. To study the relationship, grey linguistic terms were used to avoid uncertainty of data. As per our research, the decision makers have different perspectives for the data. All the factors are highly important for the achievement of Green in Lean Manufacturing for a textile industry. The result stated that C1, C4, C6, C11, and C12 are cause challenges, while C2, C3, C5, C7, C8, and C9 are effect challenges. It can also be concluded that cause from any factor very much affects the other factor.

The future scope and research include assessment of the research on the primary data and use of other methodologies.

REFERENCES

Agrawal, M., Zhou, J., & Chang, D. (2019). A survey on lightweight authenticated encryption and challenges for securing industrial IoT. In *Security and Privacy Trends in the Industrial Internet of Things* (pp. 71–94). Springer, Cham.

Arnold, C., Kiel, D., & Voigt, K. I. (2016). How the industrial internet of things changes business models in different manufacturing industries. *International Journal of Innovation Management*, 20(08), 1640015.

Boyes, H., Hallaq, B., Cunningham, J., & Watson, T. (2018). The Industrial Internet of Things (IIoT): An analysis framework. *Computers in Industry*, 101, 1–12.

Delery, J. E., & Doty, D. H. (1996). Modes of theorizing in strategic human resource management: Tests of universalistic, contingency, and configurational performance predictions. *Academy of Management JOURNAL*, 39(4), 802–835.

Hartmann, M., & Halecker, B. (2015). Management of innovation in the Industrial Internet of Things. In *ISPIM Conference Proceedings* (p. 1). The International Society for Professional Innovation Management (ISPIM), Budapest.

Kagermann, H., Lukas, W., & Wahlster, W. (2011). Industry 4.0: With the Internet of Things on the way to the 4th industrial revolution. *VDI News*, 13.

Kumar, S., & Iyer, E. (2019). An industrial IoT in engineering and manufacturing industries – benefits and challenges. *International Journal of Mechanical and Production Engineering Research and Development*, 9(2), 151–160.

Laudien, S. M., & Daxböck, B. (2016). The influence of the industrial internet of things on business model design: A qualitative-empirical analysis. *International Journal of Innovation Management*, 20(08), 1640014.

Munirathinam, S. (2020). Industry 4.0: Industrial internet of things (IIOT). In *Advances in computers* (Vol. 117, No. 1, pp. 129–164). Elsevier.

Pisano, G. P. (2015). You Need an Innovation Strategy. *Harvard Business Review*, 93(6), 44–54.

Oesterreich, T. D., & Teuteberg, F. (2016). Understanding the Implications of digitisation and automation in the context of Industry 4.0: A triangulation approach and elements of a research agenda for the construction industry. *Computers in Industry*, 83, 121–139.

Sadeghi, A. R., Wachsmann, C., & Waidner, M. (2015, June). Security and privacy challenges in Industrial Internet of Things. In *2015 52nd ACM/EDAC/IEEE Design Automation Conference (DAC)* (pp. 1–6), Piscataway, NJ.

Sengupta, J., Ruj, S., & Bit, S. D. (2020). A comprehensive survey on attacks, security issues and blockchain solutions for IoT and IIoT. *Journal of Network and Computer Applications*, 149, 102481.

Sisinni, E., Saifullah, A., Han, S., Jennehag, U., & Gidlund, M. (2018). Industrial Internet of Things: Challenges, opportunities, and directions. *IEEE Transactions on Industrial Informatics*, 14(11), 4724–4734.

Stock, T., & Seliger, G. (2016). Opportunities of sustainable manufacturing in industry 4.0. *Procedia Cirp*, 40, 536–541.

Tesch, J. F., Brillinger, A. S., & Bilgeri, D. (2017). Internet of things business model innovation and the stage-gate process: An exploratory analysis. *International Journal of Innovation Management*, 21(05), 1740002.

Thames, L., & Schaefer, D. (2016). Software-defined cloud manufacturing for Industry 4.0. *Procedia CIRP*, 52, 12–17.

Wortmann, F., & Flüchter, K. (2015). Internet of Things. *Business & Information Systems Engineering*, 57(3), 221–224.

Zhou, J., Cao, Z., Dong, X., & Vasilakos, A. V. (2017). Security and privacy for cloud-based IoT: Challenges. *IEEE Communications Magazine*, 55(1), 26–33.

8 IoT Security Issues and Solutions with Blockchain

Arvind Panwar
Guru Gobind Singh Indraprastha University

Vishal Bhatnagar
Ambedkar Institute of Advanced Communication Technologies and Research

Sapna Sinha
Amity University

Raju Ranjan
Galgotias University

CONTENTS

8.1	Introduction	142
8.2	Component of IoT	142
8.3	Issues with IoT	144
8.4	IoT Architecture and Security Challenges	146
	8.4.1 IoT Reference Model	146
	8.4.2 IoT Security Goal	148
	8.4.3 IoT Security Issue Categorization	148
	8.4.4 IoT Communication Model	151
	8.4.5 IoT Vulnerabilities	151
	8.4.6 Why IoT Needs Blockchain	153
8.5	How Blockchain Works?	154
8.6	Benefits of Blockchain-Based IoT Network	155
8.7	Different Configuration of IoT and Blockchain Integration	156
8.8	Open Challenges to Develop Blockchain-Based IoT Network	157
8.9	Conclusion	159
References		159

DOI: 10.1201/9781003140474-8

8.1 INTRODUCTION

Since the advent of World Wide Web in the 1990s, the Internet has witnessed exponential growth in diversified field. One of these fields is IoT or the Internet of Things. In today's world, our life is simplified by diverse availability of various gadgets or equipment. Linking all these gadgets through Internet and ultimately commanding them remotely through smartphone apps or other similar means will make our life more comfortable and simplified (Laroiya et al., 2020; Mohanta et al., 2020). According to an estimate, over 9 billion things are connected till date.

We can say that when physical objects are embedded with communication technology and able to sense and interact with external or internal environment, it is IoT. In other words, Internet of Things is a system of interconnected device platform that collects data and transfer data automatically using wireless technology, where devices can be any anything having the ability to collect data and transfer them on Internet like connected medical devices, biochips, sensors, or solar panels (Mistry et al., 2020).

For majority, the IoT can be understood in terms of smartphones when we connect our home appliances or gadgets with our smartphones, but the seed of IoT was germinating with the emergence of embedded system (Sengupta et al., 2020; Singh et al., 2020). When this embedded system is merged with wireless technology, control system, and home automation, it contributed to the growth of IoT. IoT is not confined to any one subject-area, rather it is a multidisciplinary field.

8.2 COMPONENT OF IoT

Figure 8.1 shows the different types of IoT component. Let's discusses them in brief.

1. **Devices or Sensors**: A sensor has basically three components namely, the sensor itself which performs sensing action and is based on some technology fit for a particular application. The second component is processing elements that convert those sensed inputs into electrical pulses. The third component is the output obtained after processing the sensed input, which is connected to the control system.

    Depending on the technology, the various sensors have been categorized as follows:
    - Inductive sensors: Inductive sensors work on electromagnetic fields which produce eddy current in metal objects.
    - Capacitive sensors: It works on measurement of capacitive change, when an object enters, the measuring field is changed which is detected by the sensor.
    - Photoelectric sensors: These sensors work on the light radiation coming to its receptor. It can measure the change in intensity of light.
    - Ultrasonic sensors: They send ultrasonic sound waves to target direction, and echo from any obstacle or object is processed to give output.
    - Magnetic field sensors: These sensors can detect any external magnetic field. The field strength can be measured to have desired output.

IoT Security Issues and Solutions 143

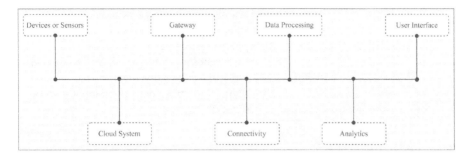

FIGURE 8.1 Component of IoT.

These are some of the examples of sensors. There are many more sensors available for different applications. These sensors play a very important role in IoT devices.

2. **Cloud System**: Cloud can be considered as a high-performance network, which interfaces various servers together and streamlines the processing in real time by many devices simultaneously. Sensors help in collecting data from different devices and applications. These data are processed efficiently and stored on the cloud system. It is also responsible for decision-making in real time (Alam, 2019; Kumar et al., 2019). Distributed database management is one of the most important components of IoT cloud. Devices, protocols, and gateways are the building blocks of IoT cloud.

3. **Gateway**: The traffic between various protocols and networks can be easily managed through gateways. It also ensures the proper connections among devices and sensors. Gateway also preprocesses the sensor data and provides encryption before sending it to the next level.

4. **Connectivity**: Sensors collect data, and the collected data must be sent to the next level for preprocessing and to store it in a database. Without connectivity, all these exercises will be futile. Hence, the connectivity is the most important aspect for IoT to work. Sensors and devices use various ways to remain connected. Wi-Fi and Bluetooth are some means to remain connected. Their benefits and disadvantages can be judged on the basis of power consumption, data transfer rate and efficiency.

5. **Data processing**: Data in IoT are usually collected through sensors and devices. These data are dynamic in nature. A proper database is needed which can store and process data efficiently that are coming from various devices and users.

6. **Analytics**: Real-time analysis of recorded data is the most important aspect of IoT technology. Analysis provides insight into any irregularities and forwards this information for action taking measures. Several enterprises are in the business of collecting bulk data and analyzing them to suggest future action or opportunity to rule the market. So, security of data and analysis are important aspects of analytics.

7. **User Interface**: The user interface is the first thing from which one can judge about the IoT system. This interface enables users to interact with

the IoT system. User-friendly and simple interfaces are more favorable than the complex one and not user-friendly. The need is to create plug and play interface to let users access the system without putting any extra effort on it. Now, there are various interface templates available which are interactive and could easily solve queries. People these days use touch panels instead of hard switches for interaction with the system. Hence, touch input is to be taken care of while designing such interfaces.

8.3 ISSUES WITH IoT

As the growth of IoT is enabling us to move toward smart homes and smart cities, it's also providing a new opportunity to adversary to intrude in our system and life. Someone even coined the term "Internet of Terror". Figure 8.2 shows the different issues with IoT.

1. **Lack of Compliance in Manufacturing Stage**: In the current scenario, millions of new IoT devices everyday are being connected to the Internet. However, many of them have serious security problems. Nevertheless, why it is happening? Because the manufacturing industry does not want to spend much time on the security layer (Hassan et al., 2019). For example, if we are talking about fitness tracking devices like fit band, they are visible on Bluetooth even after connected to one Bluetooth. These kinds of matters can be the reason for big problems. It also suggests that proper testing was not conducted at manufacturing stages. It may be a strategy to lower the cost, but later on, it creates security issues for customer and ultimately lowers the brand value. It is better to conduct all tests properly at the initial stage to save from field failure.
2. **Not Enough Updates**: according to the Gartner survey, approximately 23 billion IoT devices are connected to the network, and it can be more than 60 billion by 2025. As we can see, the total number of IoT devices is awe-inspiring. Still, developers are not doing proper testing before deployment. In many cases, later software updates can fix the problem, but many devices

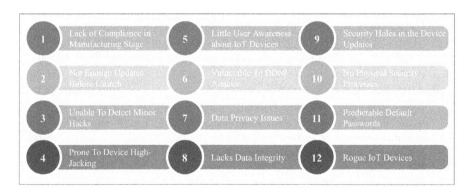

FIGURE 8.2 Issues with IoT.

are not getting software update after launching (Huang et al., 2019). Lack of proper testing before launch and lack of software updates after launch can cause a major security problem for IoT devices. Only few renowned brands take care of timely updates, but the majority of manufacturers don't care for after sale customer satisfaction.

3. **Little User Awareness**: At the starting of the 21st century, phishing and spam emailing were the cause of the big security problem, but now, the majority of Internet users understand the security risk. Though IoT is a new technology, the majority of users not wholly understand it. Thus, the little user awareness and knowledge about IoT can be a big problem for everybody in the network. Particularly in developing countries, where the majority of population is not computer literate but own such devices or phones, people don't know about the complexities of security aspects in these devices or gadgets. When people became aware of phishing and few other attacks, the adversary devise new techniques to trap individuals into their traps. By the time people became aware, they change their strategy by developing a new one.

4. **Security Holes in Update**: getting the software update is suitable, but sometimes, an unstable update can cause significant security problems. Every time during the software updating process, a device sends information to the cloud. However, if the route and files are not encrypted, then the hacker can get system information. Each update needs to be thoroughly tested before deploying. Otherwise, it may cause more problems than the earlier one.

5. **No Physical Security Processes**: So far, we have discussed about security issues related with its software, but physical security of devices is equally important. An IoT device can be hacked by the Internet as well as can be tampered physically. Instead of low-cost IoT devices, manufacturers should focus on the physical robustness of devices. It would be better to have some access control mechanism to physically access those devices.

6. **Vulnerable to DDoS Attacks**: DDoS attack is a nightmare for any network and many enterprises have faced this problem in the last two decades. Attackers always target a vulnerability in a particular network and exploit it to his benefit. IoT devices are very susceptible and exposed to DDoS attacks. An attacker can easily find such kind of devices and attack them to crash down the network (Wei et al., 2019). The IoT device manufacturers and developers need to find some solution to fight such kind of attack.

7. **Predictable Default Passwords**: It is also one of the biggest problems with manufacturers of IoT devices. Generally, they roll out many IoT devices with the same default password. Even after deployment, the user doesn't reset that default password, and as an outcome, a hacker may use that default password to hack IoT devices. In some situations, this may have a disastrous effect. For example, in healthcare and banking systems, it may cause great loss.

8. **Data Privacy Issues**: as thousands of devices are getting coupled to the Internet, a huge amount of information is generating on daily basis, and all

those data are being stored on cloud to enable the concept of smart homes and smart cities (Dorri et al., 2019; Singh et al., 2019). Besides that, when we download and install any application in our device, we agree to all terms and conditions, without even bothering what is written there. On a daily basis, IoT devices collect a huge amount of information about the user, and approximately every company buy data to get user information, and with the help of data mining tools, they try to predict user behaviors. Retailing this user data to companies is a threat to user privacy rights. The insurance company might deny my insurance renewal after assessing my driving habit collected from such applications. Manufacturers are actually invading our homes by providing connectivity through their server. Various IoT devices don't bother about user privacy rights, and user data are selling every day in market, although it is also putting a brake in the growth of IoT.

9. **Unable to Detect Minor Hacks**: Initially, the adversary launches a minor attack on IoT devices to hack minor information in background, and these kinds of minor hacks go untraceable and unnoticed. Many IoT devices are not capable to auto detect these kinds of minor hacks, and as an outcome, one day they gather all necessary information, launch a bigger attack, and ruin the system.
10. **Rogue IoT Devices**: Due to rise of phony products, there are a huge quantity of phony appliances, which exist in the market. In today's scenario, these kinds of phony IoT devices are swapping with genuine IoT devices and are installed in homes and offices. After being connected to the network and getting access of the network, they can hack or corrupt other devices in the network. The use of phony devices should be restricted to save from such situations. There should be strict standard protocols for IoT manufacturers to deal with this.
11. **Lacks Data Integrity**: As we know, IoT devices are also dealing with data transmission. IoT devices are working on data processing and data collection and sending data to the cloud. Sometimes, an IoT device can send data to cloud without encryption, and hackers can hack these data. After getting the unencrypted data, hackers can modify data and send to the cloud. In such cases, data integrity is lost (Moin et al., 2019).
12. **Prone to Device High-Jacking**: As we discussed above, IoT devices use default password, and hackers can use those passwords and can insert ransomware in the system. Ransomware is the nastiest malware; it not only snips user data but encrypts all the data and blocks data access (Sharma et al., 2019). It can lead to serious problems like home high-jacking, car high-jacking, etc.

8.4 IoT ARCHITECTURE AND SECURITY CHALLENGES

8.4.1 IoT Reference Model

Figure 8.3 shows the reference model of IoT with different layers and security solution for different layers. Reference models have seven layers in total. Let's discuss the role of different layers.

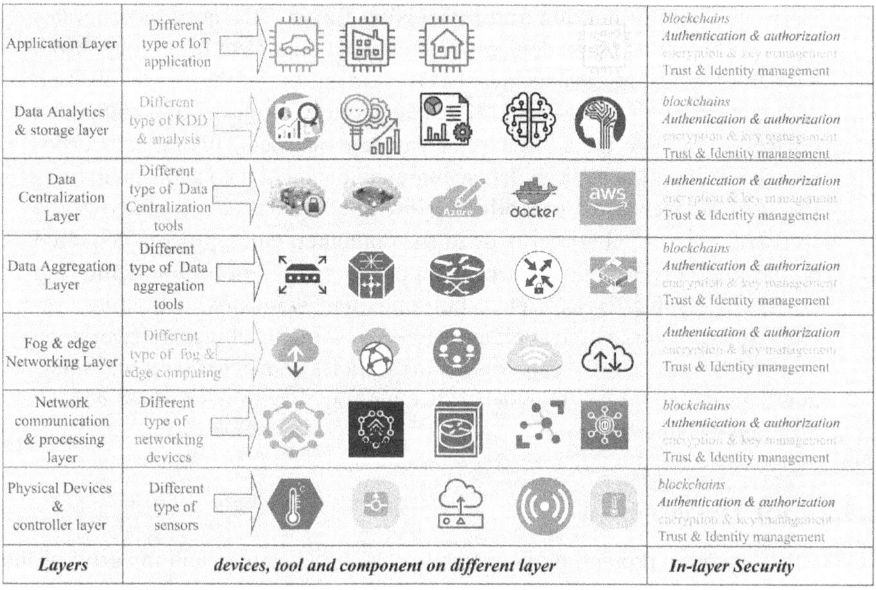

IoT Security Issues and Solutions

Layers	devices, tool and component on different layer						In-layer Security
Application Layer	Different type of IoT application						blockchains Authentication & authorization encryption & key management Trust & Identity management
Data Analytics & storage layer	Different type of KDD & analysis						blockchains Authentication & authorization encryption & key management Trust & Identity management
Data Centralization Layer	Different type of Data Centralization tools						Authentication & authorization encryption & key management Trust & Identity management
Data Aggregation Layer	Different type of Data aggregation tools						blockchains Authentication & authorization encryption & key management Trust & Identity management
Fog & edge Networking Layer	Different type of fog & edge computing						Authentication & authorization encryption & key management Trust & Identity management
Network communication & processing layer	Different type of networking devices						blockchains Authentication & authorization encryption & key management Trust & Identity management
Physical Devices & controller layer	Different type of sensors						blockchains Authentication & authorization encryption & key management Trust & Identity management

FIGURE 8.3 Reference model of IoT.

1. **Application Layer**: This layer belongs to the end user to host different vertical and horizontal applications for various area. There are lots of application areas like healthcare, smart education, smart power grid, smart cities, and so on.
2. **Data Analysis and Storage Layer**: This layer is designed to analyze the data from different sources and store different insights after analysis. This layer also defines application-based analytics engines like analytics for e-health, analytics for smart cities, and so on. This layer uses complete functions of knowledge discovery in data.
3. **Data Centralization Layer**: This layer performs the task of data centralization. This resembles the essential networking jobs of up-to-date networks. It comprises the functionality of characteristically originating industry-specific extranets, enterprise-owned networks, Internet tunnels, and hybrid/private/public cloud-oriented connection. These networks complete their task by employing carrier-provided amenities and use wire or wireless connection.
4. **Data Aggregation Layer**: This layer performs data aggregation job that includes some kind of summarization of data or conversion of protocol. This data aggregation job is classically held in an entryway device.
5. **Fog and Edge Computing Layer**: This layer has the provisions of edge computing and fog networking, that is, the neighborhood-specific network which is the initial stage of the IoT client connectivity. Characteristically, fog computing is improved to the IoT clients' use-dedicated protocols. It can be wired link or wireless connection. This layer cares the early communication connection used by devices, for example, cellular links, Bluetooth Low Energy (BLE), ZigBee, Wi-Fi, and so on.

6. **Network Communication and Processing Layer**: This layer incorporates different communication devices which collect data from things or sensors from the physical layer. This layer has the capabilities of the data acquisition and data processing. It is actually connected to the sensors and embedded devices and sensor hubs via wire or wireless Internet. This layer incorporates local attention devices that accumulate the local data and signals.
7. **Physical devices and Controller Layer**: This layer is encompassed of the world of "objects" that are focus to the computerization process delivered by the IoT. This is a huge sphere, with people (with e/m-health monitoring devices, wearable devices, etc.), buildings and homes (with lighting systems, smart meter), home appliances (like washing machines, refrigerators, air conditioners), CCTV cameras, smart vehicles connected with IoT, smart power grid, and so on. This layer also has some controllers which control different devices.

8.4.2 IoT Security Goal

IoT security involves protecting the reliability, usability, safety, and integrity of the IoT network and IoT data. Actual IoT security downfalls a variability of threats from incoming or distribution on an IoT network. The prime goals of IoT security are integrity, confidentiality, and availability. These three supporting pillars of IoT security are frequently characterized as CIA trio. Figure 8.4 shows the IoT security goals.

- **Integrity**: Assuring and maintaining the consistency and accurateness of data. The purpose of integrity is to assure that the data in IoT are changed by only authorized individuals.
- **Confidentiality**: The meaning of confidentiality is to shield valuable data from unauthorized folks. Confidentiality assures that the data are accessible only to the authorized and intended people.
- **Availability**: The purpose of availability in IoT security is to assure that the IoT resources, IoT services, and IoT data are uninterruptedly accessible to the authentic workers.

8.4.3 IoT Security Issue Categorization

IoT architype incorporates a very wide range of devices and apparatus oscillating from the high-end computer server to very small size embedded chips. This requires to point out different security issues at various levels. A categorization of different security issues at different level is shown in Figure 8.5. The figure shows three different levels according to IoT development.

1. **Low-level Security Issue**: The first level of security issues is low-level security issue which belongs to physical layer and data communication layer. This kind of security issue is also known as hardware level issues. Some of them are described below:

IoT Security Issues and Solutions 149

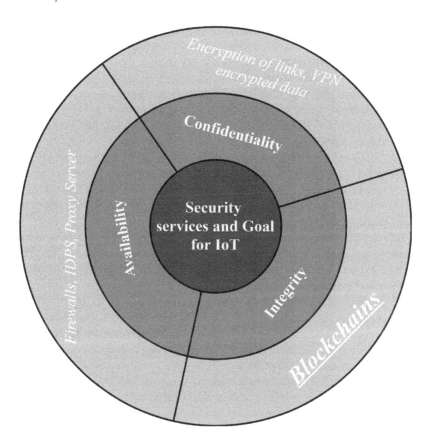

FIGURE 8.4 Security goal of IoT.

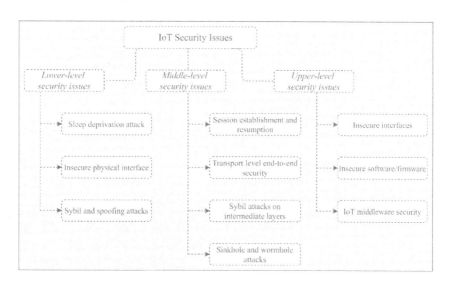

FIGURE 8.5 IoT security issues.

- Sleep deprivation attack: In this, victim node keeps on getting request time to time so that it cannot get into low-energy sleep mode. This makes victim node awake, hence preventing them from performing huge data-intensive tasks.
- Insecure physical interface: Numerous physical aspects pose multiple serious threats to suitable working of sensors in IoT. The poor physical interfaces may be demoralized to compromise security in the network.
- Low-level sybil and spoofing attacks: This kind of attacks in an IoT network is triggered by nasty sybil nodes, who practice forged personalities to damage the IoT working. A sybil node at physical layer may use fake MAC address for camouflage as a diverse device. Subsequently, the genuine nodes denied contact to resources.

2. **Intermediate-Level Security Issues**: This kind of security issues is primarily focused on communication, session management, and routing functions happening at network or transport layers of IoT. Some of them are described below.
    - Session establishment and resumption: In session hijacking counterfeit messages leads to denial of service (DOS) attack, in this malicious node pretends to be victim node and takes hold of the session. It happens at transport layer; communicating nodes alter the sequence number to initiate re-transmission of data.
    - Transport level end-to-end security: To ensure end-to-end security at transport layer, so that reliable delivery of the message sent by sender is there to the destination.
    - Sybil attacks on intermediate layers: In IoT, sybil nodes are responsible to initiate sybil attack by creating multiple subverted fake identities to damage the IoT working. A sybil node at physical layer may use fake MAC address for being masked as a diverse device.
    - Wormhole and Sinkhole attacks: In warm hole attacks, the nodes create a tunnel between them for direct delivery and creating shortcut path in wireless network. It is also related to sinkhole node where convincing nodes can advertise itself as sink node due to which all the traffic gets diverted to the same route toward sink node.

3. **High-Level Security Issues**: This kind of security problems are primarily focused with the applications hosted on IoT. Some of them are defined below.
    - Insecure interfaces: To use IoT services, different interfaces are used over cloud, mobile, and web that are susceptible to diverse security attacks which may brutally disturb the data secrecy.
    - Insecure firmware: Several susceptibilities in IoT contain those triggered by uncertain firmware and software. The code written if different languages like XML, JSON, XSS, and SQLite require proper testing before deployment. Equally, the firmware and software updates to be approved in a very secure way.

IoT Security Issues and Solutions

- **Middleware security**: The IoT planned to concentrate on messaging amid mixed objects of the IoT architype must be secure. Diverse environments and interfaces used by middleware must deliver secure communication.

8.4.4 IoT Communication Model

In IoT networks, devices like different types of sensors are connected to each other and the Internet. These devices are responsible to keep on sensing and collecting data and end it to data storage arrays on cloud. Network devices like router and gateways allow devices to connect and interact via Wi-Fi. Data analysis and controlling of IoT devices is done on a cloud by the user using remote devices. Figure 8.6 explains the communication model of IoT networks.

8.4.5 IoT Vulnerabilities

There are various vulnerabilities existing in current IoT networks and infrastructure; before understanding how blockchain can help in securing the IoT networks, understanding of vulnerabilities is expected. The vulnerabilities are shown in Figure 8.7 and explained below:

- **Obsolete Firmware**: With the frequent advancements, the firmware needed to be updated, but it is found that many manufacturers do not release updates and even users do not bother to update the firmware. Due to which obsolete firmware becomes more susceptible to malwares and reduced performance.

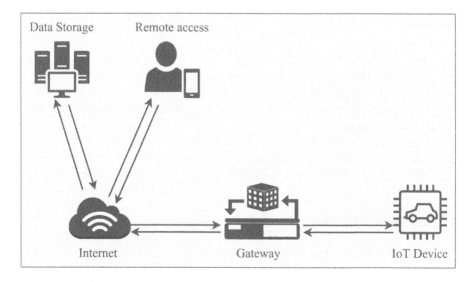

FIGURE 8.6 IoT communication model.

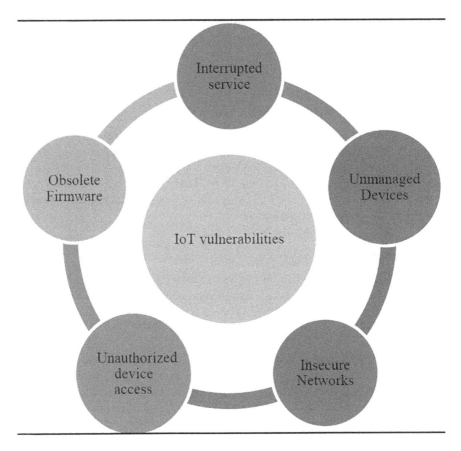

FIGURE 8.7 IoT vulnerabilities.

- **Unauthorized Device Access**: Weak authentication results in guessable passwords. As all devices are connected in IoT network, it is very easy for the hacker to penetrate into the network from any device by using or guessing the password.
- **Insecure Networks**: Network implementation is an ongoing process due to use of different devices, protocols, and standards leaving gaps which provide easy access to hackers. Data leaks are the usual phenomena observed in transferring of data between device to device or device to storage if it is not encrypted.
- **Unmanaged Devices**: Due to lack of professional device management, IoT Networks do not perform well and get more vulnerable to cyber-attack.
- **Interrupted Service**: Physical damage of device, network failure, and security issues of cloud are the other important factors. Cyber attackers can reconfigure the devices and manipulate the networks by initiating DOS attack for service interruption.

IoT Security Issues and Solutions

8.4.6 WHY IoT NEEDS BLOCKCHAIN

There are many factors which prove that IoT needs blockchain, like:

- Blockchain adds trust to the data received from IoT devices.
- Blockchain makes IoT devices secure.
- Food tracking is a popular example for blockchain and IoT application.
- Blockchain is an online spreadsheet associated with bitcoin and other digital currency.

The same way miners create "Blocks" by solving complex math problems, cryptocurrency bit coin is based on the blockchain technology, in which blocks are created for each transaction (Viriyasitavat et al., 2019). All the blocks are connected together forming a chain of blocks. These blocks are timestamped and unalterable. Any change in block leads it to get verified by all the parties in the trusted domain, and hash value is also changed. The hash value of the previous block is stored in a block, and hence, change in content makes block out of the chain. There are some key factors of blockchain like it is secure, transparent, easy, and low cost.

What is blockchain in IoT? IoT is a network, connected to Internet using wireless technology of smart computers, objects and devices that gather and share enormous volumes of data. When we integrate both blockchain and IoT, it becomes a blockchain of things (BCoT). BCoT consists of the layers from bottom to top perception layer, communication layer, blockchain composite layer, and industrial application (Minoli & Occhiogrosso, 2018). The main features of BCoT are as follows:

- Blockchain works between IoT and industrial applications.
- It offers an abstraction in IoT at lower layer and provides blockchain-based services.
- It hides the complexity of IoT.
- It offers general interfaces to various IoT applications.

BCoT brings many opportunities:

- Enhanced interoperability of IoT systems.
- Improved security of IoT systems.
- Traceability and reliability of data in IoT networks.
- Autonomic interactions of IoT systems.

There is a wide diversity of BCoT including smart manufacturing, supply chain management, food industry, smart grid, and healthcare Internet of vehicles.

The blockchain is a protected conveyed information base for recording exchanges. With blockchain, the estimation of any sort can be traded and straightforwardly shared without a concentrated power to record and approve the exchange. The blockchain can give an automatic and secure exchange framework for cutting-edge IoT systems. Its ability to deal with billions of associated gadgets empowers the frictionless handling of exchanges and coordination between gadgets. This decentralized methodology could take out single purposes of disappointment and bottlenecks and make information and exchanges more secure (Reyna et al., 2018). The example of blockchain in IoT is "Slock" which is home automation application. An entryway lock is associated with a keen agreement on the blockchain that controls when and who can open the lock. Such innovation connects the blockchain and the physical world by making mechanized keen agreements, consequently empowering the frictionless and computerized brief trade of benefits, for example, bicycles, parking garages, or rooms. Utilizing blockchain for IoT offers better approaches to robotize business cycles and manufacture appropriate independent assistance frameworks (Khan et al., 2018; Nguyen & Kim, 2018).

8.5 HOW BLOCKCHAIN WORKS?

The basic concept behind blockchain is distributed ledger. When any transaction is initiated, a block is created which contains information about the transaction, and then that block is broadcasted to all the parties/nodes on the network for the consensus. When all the parties/nodes validate and give their consensus, the block is added into the chain, and transaction is executed. The beauty of blockchain is that each block in the chain is secure enough so that if any change in the block happens, the hash value calculated on the block changes. For adding it again in the chain requires consensus from parties again, and hence, these blocks are secure for any change and manipulation. Figure 8.8 shows the steps used in blockchain.

To summarize, the blockchain can improve the issue of privacy and security existing in IoT systems by:

- Not allowing modification and change in the blocks containing transaction data.
- Any modification in the block is not possible; only changed block is allowed to add into the chain.
- These blocks are timestamped so history of the data can be traced easily.

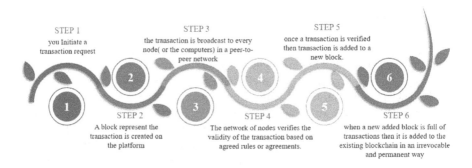

FIGURE 8.8 Working of blockchain.

IoT Security Issues and Solutions

8.6 BENEFITS OF BLOCKCHAIN-BASED IoT NETWORK

Figure 8.9 shows the benefits of blockchain-based IoT network.

- **Data Decentralization**: The core concept behind blockchain technology eliminates the issues raised in IoT network. IoT networks have centralized data processing; data are stored on centralized servers or on clouds. Centralized processing makes data vulnerable at a single point, and if centralized data are manipulated or fail, whole network will get collapsed (Qian et al., 2018). Blockchain architecture prevents a single point of failure by decentralization of process, cost for infrastructure, and decentralization management, improving fault tolerance.
- **Improved System Scalability**: Decentralization also enhances scalability, and distributed workloads on many nodes help in increasing computing power. The problem of having additional storage, bandwidth, and compute power for scalability in IoT can easily be solved using blockchain.
- **Guaranteed Data Immutability**: Blockchain prevents data to be changed using cryptography, and this ensures secure data transfer and storage. This feature of blockchain allow nonchangeable history of smart device communications in IoT network.

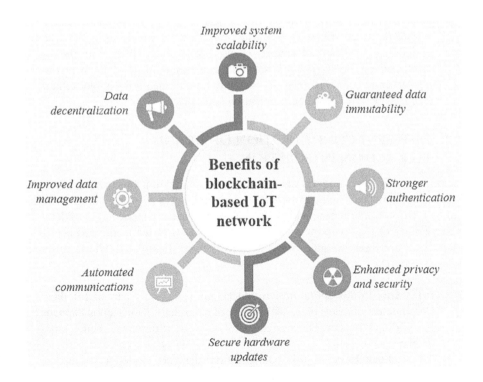

FIGURE 8.9 Benefits of blockchain-based IoT network.

- **Stronger Authentication**: Blockchain ensures trusted authentication and authorization of devices. Blockchain uses decentralized approach of Public Key Infrastructure (PKI) for generating public and private key pair that is used for authenticating devices at the time of registration (Dorri et al., 2017; Mendez Mena et al., 2018). Man-in-the-middle attack can easily counter with the use of approach used in blockchain.
- **Enhanced Privacy and Security**: Blockchain secures data during transition by storing it in blocks which is unchangeable, and use of cryptography while communication also eliminates the information breaches at end-to-end.
- **Secure Hardware Updates**: Since blockchains are secure and unchangeable, it allows developer to push code to IoT devices securely, making updating of software of devices easy. Two sublayers of blockchain namely infrastructure and smart contract allow support of multiple devices and services. Where infrastructure sublayer manages certificate authority, consensus protocol, and unchangeable distributed ledger, smart contract verifies firmware and its updates.
- **Automated Communications**: Data gathered by devices can be treated as transactions similar to blockchain, and using smart contracts, each transaction can be automated. Use of smart contract in IoT devices makes it possible to conduct automatic machine-to-machine transaction without third-party approval.
- **Improved Data Management**: In IoT networks, a huge amount of data is transferred to central storage residing on the cloud which raises the issue of data management and security (Kumar & Mallick, 2018). With the use of blockchain, devices of IoT networks can communicate without involving server and cloud storage. As there are no intermediaries in blockchain, encrypted peer-to-peer communication takes place.

8.7 DIFFERENT CONFIGURATION OF IoT AND BLOCKCHAIN INTEGRATION

There are different configurations available for integrating blockchain with IoT. Implementation of any configuration starts with creation of network. These networks can be IoT network or blockchain or combination of both. The second consideration is availability of fog computing sub layer. On the basis of the mentioned consideration, there are three possible configurations defined. Figure 8.10 shows different types of integration.

- **IoT – IoT**: In this configuration, blockchain stores only part of IoT data. Communication takes place without use of blockchain. Transaction happens fast with low latency. Devices can also work on offline mode, hence more secure.
- **IoT – Blockchain**: In this configuration, data are shared between IoT devices using blockchain, which act similar to cloud used in IoT network.

IoT Security Issues and Solutions

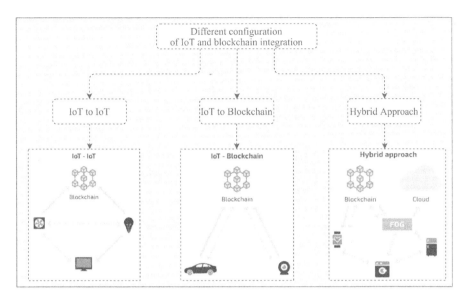

FIGURE 8.10 Different types of configurations.

Due to blockchain autonomy of devices, security and traceability of communication increase along with increased bandwidth.

- **Hybrid Approach**: In this approach, data transfer happens directly among IoT devices and blockchain stores only partial data leveraging benefits of real-time IoT and blockchain. Fog and cloud computing are implemented to overcome the limitation of both blockchain and IoT devices. Handling of data at edge devices reduces the operating expenses.

8.8 OPEN CHALLENGES TO DEVELOP BLOCKCHAIN-BASED IoT NETWORK

There are different technical challenges associated with the integration of IoT and blockchain. This section deals with the various challenges and their solutions. Figure 8.11 shows some challenges.

- **Difficulties in Choosing a Consensus Protocol**: For integration of blockchain with IoT, the major issues addressed by the developers are expensive nature of blockchain computation and resource constraint of IoT devices. For addressing these issues, a consensus protocol for IoT network was developed. The consensus protocol defines the methods used by different nodes to reach to consensus of over new node. The consensus protocol used in blockchain is not found suitable for IoT networks due to high computation and cost (Banafa, 2017; Lamba et al., 2017). There is a need of consensus algorithm that can consider requirements of IoT yet to be proposed.

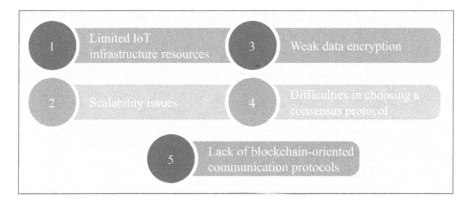

FIGURE 8.11 Open challenges to develop blockchain-based IoT network.

- **Limited IoT Infrastructure Resources**: The space required in blockchain computation is more than the available resources in IoT networks. IoT devices have limited computational, memory, and storage capacity. To address the challenge of blockchain size, IBM has proposed categorization of nodes as light peers, standard peers, and peer exchange. These nodes work as per their storage, networking, power, and processing capabilities. Information about own blockchain address and balance is stored by light nodes, recent transactions done by the node and light peers are stored on standard peers, and storing complete blockchain and data analysis is done by peer exchanges (Kshetri, 2017; Park & Park, 2017).
- **Weak Data Encryption**: Encryption is used in blockchain for data security, but in IoT, devices interact with each other automatically, and hence, reliable cryptography cannot be achieved. The possible solution is by providing entropy by quantum random number generator leveraging quantum physics.
- **Scalability Issues**: The biggest challenge in integration of blockchain and IoT technologies is scalability. There is continuous addition of smart devices in the IoT networks, which leads to more data generation and transactions. IoT devices work in real time, which is another issue in integration of both the technology where blockchain needs time for confirmation of transactions (Dorri et al., 2016). The blockchain architecture that can process high throughput of transaction along with security of peers is required.
- **Lack of Blockchain-Oriented Communication Protocols**: A standard security protocol that can fit in all types of IoT devices and IoT networks needed to be developed. Diversity in devices affects security and privacy and its integration with blockchain. The use of Wi-Fi in communication in IoT networks is not suitable for blockchains (Ali et al., 2015; Nandan Kumar & Vajpayee, 2016). The IoT networks are generally application-specific that often lack operating systems, whereas in conventional network, the same operating system is used. Another issue in traditional IoT networks end devices gets connected to Internet or gateway device using secure and fast

connection, but smart devices are connected through low bandwidth and power Wi-Fi communication. The suitable protocol which can solve all the issues is needed.

8.9 CONCLUSION

In today's world, our life is simplified by diverse availability of various gadgets or equipment. Linking all these gadgets through Internet and ultimately commanding it remotely through smartphone apps or other similar means will make our life more comfortable and simplified. On the arrival of smart cities, smart power grids, smart homes, and the whole smart things around us, the IoT (Internet of Things) has arisen as a field of unbelievable potential, impact, and exponential growth. According to Cisco Inc. prediction, approx. 50 billion devices will be connected to the Internet by the end of 2020. This chapter presented a survey on IoT and different security issues in IoT. The authors provide a categorization of security issues in three levels (low, intermediate, and high). After categorization of security issues, the authors present working of blockchain technology with brief introduction. At the end, most importantly, the authors present how blockchain can play a vital role to resolve various IoT security issues.

REFERENCES

Alam, T. (2019). Blockchain and its role in the Internet of Things (IoT), 5(1). arXiv. doi:10.32628/cseit195137.
Ali, M., Khan, S. U., & Vasilakos, A. V. (2015). Security in cloud computing—Opportunities and challenges. *Information Sciences*, 305, 357–383. doi:10.1016/j.ins.2015.01.025.
Banafa, A. (2017). IoT and blockchain convergence—Benefits and challenges. *IEEE Internet of Things*, January 10. http://iot.ieee.org/newsletter/january-2017/iot-and-blockchain-convergence-benefits-and-challenges.html.
Dorri, A., Kanhere, S. S., & Jurdak, R. (2016). Blockchain in internet of things—Challenges and solutions. Arxiv.Org. http://arxiv.org/abs/1608.05187.
Dorri, A., Kanhere, S. S., Jurdak, R., & Gauravaram, P. (2019). LSB—A lightweight scalable blockchain for IoT security and anonymity. *Journal of Parallel and Distributed Computing*, 134, 180–197. doi:10.1016/j.jpdc.2019.08.005.
Dorri, A., Kanhere, S. S., Jurdak, R., & Gauravaram, P. (2017). Blockchain for IoT security and privacy—The case study of a smart home. *2017 IEEE International Conference on Pervasive Computing and Communications Workshops, PerCom Workshops 2017*, pp. 618–623. doi:10.1109/PERCOMW.2017.7917634.
Hassan, M. U., Rehmani, M. H., & Chen, J. (2019). Privacy preservation in blockchain based IoT systems—Integration issues, prospects, challenges, and future research directions. *Future Generation Computer Systems*, 97, 512–529. doi:10.1016/j.future.2019.02.060.
Huang, J., Kong, L., Chen, G., Wu, M. Y., Liu, X., & Zeng, P. (2019). Towards secure industrial iot—Blockchain system with credit-based consensus mechanism. *IEEE Transactions on Industrial Informatics*, 15(6), 3680–3689. doi:10.1109/TII.2019.2903342.
Khan, M. A., Salah B A Bahauddin, K., Multan, Z. U., & Khalifa, P. B. (2018). IoT security—Review, blockchain solutions, and open challenges. *Future Generation Computer Systems*, 82, 395–411. doi:10.1016/j.future.2017.11.022.
Kshetri, N. (2017). Can blockchain strengthen the Internet of Things? *IT Professional*, 19(4), 68–72. doi:10.1109/MITP.2017.3051335.

Kumar, K. D., Venkata Rathnam T., Venkata Ramana R., Sudhakara, M., & Poluru, R. K. (2019). Towards the integration of blockchain and IoT for security challenges in IoT. *igi-global.com* (pp. 45–67). doi:10.4018/978-1-7998-0186-3.ch003.

Kumar, N. M., & Mallick, P. K. (2018). Blockchain technology for security issues and challenges in IoT. *Procedia Computer Science, 132*, 1815–1823. doi:10.1016/j.procs.2018.05.140.

Lamba, A., Singh, S., Singh, B., Dutta, N., Sai, S., & Muni, R. (2017). Mitigating IoT security and privacy challenges using distributed ledger based blockchain (DL-BC) technology. *International Journal for Technological Research in Engineering, 4*(8). https://www.ijtre.com.

Laroiya, C., Saxena, D., & Komalavalli, C. (2020). Applications of blockchain technology. *Handbook of Research on Blockchain Technology, 21*(3), 213–243. doi:10.1016/b978-0-12-819816-2.00009-5.

Mendez Mena, D., Papapanagiotou, I., & Yang, B. (2018). Internet of things—Survey on security. *Information Security Journal, 27*(3), 162–182. doi:10.1080/19393555.2018.1458258.

Minoli, D., & Occhiogrosso, B. (2018). Blockchain mechanisms for IoT security. *Internet of Things, 1–2*, 1–13. doi:10.1016/j.iot.2018.05.002.

Mistry, I., Tanwar, S., Tyagi, S., & Kumar, N. (2020). Blockchain for 5G-enabled IoT for industrial automation—A systematic review, solutions, and challenges. *Mechanical Systems and Signal Processing, 135*. doi:10.1016/j.ymssp.2019.106382.

Mohanta, B. K., Jena, D., Satapathy, U., & Patnaik, S. (2020). Survey on IoT security—Challenges and solution using machine learning, artificial intelligence and blockchain technology. *Internet of Things, 11*, 100227. doi:10.1016/j.iot.2020.100227.

Moin, S., Karim, A., Safdar, Z., Safdar, K., Ahmed, E., & Imran, M. (2019). Securing IoTs in distributed blockchain—Analysis, requirements and open issues. *Future Generation Computer Systems, 100*, 325–343. doi:10.1016/j.future.2019.05.023.

Nandan Kumar, S., & Vajpayee, A. (2016). A survey on secure cloud—Security and privacy in cloud computing. *American Journal of Systems and Software, 4*(1), 14–26. doi:10.12691/ajss-4-1-2.

Nguyen, G. T., & Kim, K. (2018). A survey about consensus algorithms used in blockchain. *Journal of Information Processing Systems, 14*(1), 101–128. doi:10.3745/JIPS.01.0024.

Park, J. H., & Park, J. H. (2017). Blockchain security in cloud computing—Use cases, challenges, and solutions. *Symmetry, 9*(8), 1–13. doi:10.3390/sym9080164.

Qian, Y., Jiang, Y., Chen, J., Zhang, Y., Song, J., Zhou, M., & Pustišek, M. (2018). Towards decentralized IoT security enhancement—A blockchain approach. *Computers and Electrical Engineering, 72*, 266–273. doi:10.1016/j.compeleceng.2018.08.021.

Reyna, A., Martín, C., Chen, J., Soler, E., & Díaz, M. (2018). On blockchain and its integration with IoT. Challenges and opportunities. *Future Generation Computer Systems, 88*, 173–190. doi:10.1016/j.future.2018.05.046.

Sengupta, J., Ruj, S., & Das Bit, S. (2020). A comprehensive survey on attacks, security issues and blockchain solutions for IoT and IIoT. *Journal of Network and Computer Applications, 149*, 102481. doi:10.1016/j.jnca.2019.102481.

Sharma, T., Satija, S., & Bhushan, B. (2019). Unifying blockchain and IoT: Security requirements, challenges, applications and future trends. *Proceedings 2019 International Conference on Computing, Communication, and Intelligent Systems, ICCCIS 2019*, January, pp. 341–346. doi:10.1109/ICCCIS48478.2019.8974552.

Singh, S. K., Rathore, S., & Park, J. H. (2020). BlockIoTIntelligence—A blockchain-enabled intelligent IoT architecture with artificial intelligence. *Future Generation Computer Systems, 110*(8), 721–743. doi:10.1016/j.future.2019.09.002.

Singh, S. K., Singh, S. K., Rathore, S., Park, J. H., Rathore, S., & Park, J. H. (2019). BlockIoTIntelligence—A blockchain-enabled intelligent IoT architecture with artificial intelligence blockchain-based secure storage management with edge computing for IoT View project social network security view project BlockIoTIntelligence—A Blockchai. Elsevier. doi:10.1016/j.future.2019.09.002.

Viriyasitavat, W., Anuphaptrirong, T., & Hoonsopon, D. (2019). When blockchain meets Internet of Things—Characteristics, challenges, and business opportunities. *Journal of Industrial Information Integration*, *15*, 21–28. doi:10.1016/j.jii.2019.05.002.

Wei, L., Wu, J., Long, C., & Lin, Y. B. (2019). The convergence of IoE and blockchain—Security challenges. *IT Professional*, *21*(5), 26–32. doi:10.1109/MITP.2019.2923602.

9 Stabilization of Imbalance between the Naira and the Dollar Using Game Theory and Machine Learning Techniques

Garba Aliyu and Bukhari Badamasi
Ahmadu Bello University

Sandip Rakshit and Onawola, H.J.
American University of Nigeria

CONTENTS

9.1 Introduction 163
    9.1.1 General Understanding 164
        9.1.1.1 Foreign Exchange 164
        9.1.1.2 Game Theory 165
        9.1.1.3 Machine Learning 165
9.2 Literature Review 166
9.3 Data Understanding/Problem Statement 168
9.4 Conceptual Framework 170
9.5 Conceptual Model 171
9.6 Conclusion 171
References 172

9.1 INTRODUCTION

Nigeria is a country that has the biggest economy in Africa, even though it relies solely on foreign currency, particularly US dollars. Consequently, this makes the Naira have no significant value over dollars. The central bank has for quite a long time attempted to keep the Naira stable at around 315 to the Dollar, after apparently drifting it last June of 2017, however a deficiency of remote cash joined with appeal for dollars has made the Naira lose as much as 38% of its value on the black market from that point forward (Economist, 2017). The conflicting between the Naira and Dollar is significant for study to stabilize the Naira. The contemporary issue is having Naira value to be at the high side, there may be consequences from the investors' side. There should be a strike balance between the Naira and Dollar. However, many factors influence the high value of Dollar as against Naira. Example of such include importation of many things from other countries. This is coming because of dilapidation of most of our manufacturing industries and lack of establishment of new ones, which results to high rate of unemployment among youths. Accuracy in forecasting foreign currency exchange rates, or at the very least predicting the trend, is critical for effective investments in today's competitive global economy. In recent years, the use of computational intelligence methods for forecasting has proven to be extremely effective (Papatsimpas et al., 2020).

Game theory is simply the numerical hypothesis of key cooperation between self-intrigued operators. Game theory gives a scope of models to speaking to vital collaborations and related to these, a group of arrangement ideas, which endeavor to describe the normal results of games. Game theory is essential to software engineering for a few reasons: First, communication is a major point in software engineering, and if it is expected that framework segments are self-intrigued, at that point the models and arrangement ideas of game theory appears to give a fitting structure which to display such frameworks. Second, the issue of processing with the arrangement ideas proposed by game theory raises significant difficulties for software engineering, which tests the limits of current algorithmic strategies. A number of variables complicates interactions with an opponent. First, determining the consequences of an agent's behavior can be difficult due to hidden knowledge or because the effects are partly dependent on the opponent's parallel action (Damer et al., 2019). This research plans to present the key ideas of game hypothesis for a software engineering crowd, underscoring both the materialness of game-theoretic ideas in a computational setting and the job of computation in game-theoretic issues to address the imbalance of Naira and Dollar. Machine learning techniques will be used to predict the appreciation or depreciation of Naira upon policy (rules) put in place by the government, which is to be guided by gross domestic product (GDP).

9.1.1 General Understanding

9.1.1.1 Foreign Exchange

The foreign exchange market is a place where you can trade and exchange any currency pair. The 'foreign exchange rate' is the value (price) of one currency in terms of another currency. Demand and supply for currencies over time, based on trade value, capital flows, and consumer expectations, decide exchange rate movements. Spot, Forwards, Swaps, Options, and Futures are all available on the Nigerian foreign exchange market. According to FMDQ Securities Exchange (2020), the most exchanged currency pair in the market is the USD/Naira ($/N), but market makers also deal in crosses such as Canadian Dollars (CAD), Swiss Francs (CHF), Euros (EUR), Pound Sterling (GBP), Japanese Yen (JPY), and South African Rand (ZAR). The market forces that drive foreign exchange rates can be influenced by a variety of factors. Various economic, political, and even psychological conditions are among the influences. Economic policies, trade balances, inflation, and the prospects for economic growth are all factors to consider. According to CFI (2020), many things such as political instability and political conflicts can have a negative effect on a currency's value, and this has a major impact on the foreign exchange rate. Exchange rates can also be influenced by the psychology of foreign exchange market participants.

9.1.1.2 Game Theory

There are certain phenomena where two or more players are involved in decision-making under certain conflict circumstances. In this case, the right decision is best achieved by maximizing the benefits and minimizing the losses. The decision depends on the decision variables chosen by the players. Hence, game theory is a concept that provides the best way to make a decision when two or more intelligent and discerning players are involved under contingencies of conflict. Game theory could be combined with the multiagent approach to deal with many empirical situations. Stefano et al. (2020) believe that using a game-theoretic method combined with heuristics to validate the effectiveness of the modeling approach for the two blockchain conflict scenarios using actual data sets would be extremely beneficial. Multiplayer is the two sides of the game in the same game stage. It is critical to figure out how to increase both parties' interests (Ma, 2020). Thus, in our model, we considered Naira and Dollar as two conflicting entities, competitors, or players. Moreover, the theory of games started as far back as the 20th century even though in 19194, John Von Neumann and Morgenstern published a paper titled "Theory of Games and Economic Behavior", which conceptualized the theory using a mathematical concept (Rama, 2007). Decision-making requires placing the different prospects and economically assessing to select the best among them. Noncooperative game theory was essentially interested in the coordination of different operators and formally represented several equilibrium assumptions whose main characteristics were specifically scrutinized. These assumptions transpired from the modeler's point of view, signifying inevitable circumstances ensuring compatibility between expectations and strategies for every player, amid plans for all players. The modeler displays concrete coordination rules by which they arrive at adjusting their respective actions to attain a balanced state (Routledge, 2004).

9.1.1.3 Machine Learning

Machine learning is a field of study that allows a computer to learn and deduce from data without being explicitly programmed. This can be best described as given task T, a computer expects to learn from T to gather an experience E and measure the performance P, if the measured performance by P improves based on the experience E gathered on the task T (Aurélien, 2019). A collection of values of the games (saddle points) will be used to train the machine learning model, and test and validate its efficiency on how best it can forecast the possible values in the future. However, more data to train is better for the model to forecast accurately from unseen data (test data). A large dataset enables the algorithm to learn and gather more experience and intelligently predict accurately when new unseen data are encountered. Achieving high accuracy with a large dataset may be at the expense of time, even though dimensionality reduction helps to improve the performance (Isma'il et al., 2020). The form of machine learning that is currently being studied in the most recent research is supervised learning. It entails learning from a collection of labelled examples given a qualified external overseer as part of a training data set. The labelled examples form the training data make up of description and specifications – the label of the ideal behavior the system should take in a given situation. The aim of the supervised learning is to create a system that can extrapolate and generalize its responses in order to make correct decisions (Usha et al., 2019). Since the predicted performance is either a rise or fall in the foreign currency rate trend, Forex prediction problems are commonly referred to as binary classification problems (Thu & Xuan, 2018). A hybrid form of artificial intelligence has recently been found to be able to estimate the 30-minute time limit. This innovation enables traders to benefit within the time frame by predicting all price indicators such as open, near, big, and low (Sidehabi & Tandungan, 2016).

9.2 LITERATURE REVIEW

Most of the problems associated with machine learning that has to do with many-objective optimization problems could be translated in a way that two or more conflicting objectives resulted in a more optimized solution. The multioptimization problems could be mapped to game theory to obtain stable solutions (Rekha et al., 2016). Game theory is a field in the operation research that can be applied to the different ambit of artificial intelligence (AI) (Ippolito, 2019). If a Game theory is combined with AI, modeling an AI system that comprises of many agents which will collaborate and compete to achieve a specific objective. This is a situation where game theory plays an important role. Stefano et al. (2020) modeled two conflict situations in their work and suggested a novel approach focused on game theory and multilayer networks. Besides, the work has included human-related variables that influence the agents' decisions, such as homophily. We presently talk about some comparable circumstances that emerge in zero-sum games. A zero-sum game is a mathematical portrayal of a circumstance where the total of benefits (might be positive or negative) of the considerable number of players is equivalent to zero (Agrawal & Jaiswal, 2012). Game theory has concentrated on modeling the most well-known collaboration designs that now we are seeing each day in multiagent AI systems. Understanding the various sorts of game elements in a situation is a key component to structure productive gamified AI systems. At a significant level, the game elements in AI systems has a five-component criteria as depicted in Figure 9.1.

Game theory was used by Serdar (2012) to analyze the international relation conflict between Iran and Israel, where two strategies were assumed for each of the countries. The two strategies considered for Iran are "stop nuclear research" and "do not stop", while Israel has "attack" and "do not attack", thus forming a matrix of 2 × 2 strategies and two players. The findings revealed that regardless of Iran's choices, Israel has better outcomes by adopting "do not attack". Conversely, regardless of Israel's choices, Iran has better outcomes by adopting "continue nuclear research". Moreover, the equilibrium is to "continue nuclear research" and "do not attack". To bring balance about the contenders, Ma (2020) considered establishing a new model to optimize overall interests so that all players in the game can have a good experience. A game called "crossing the desert" was used to allow players to choose the

FIGURE 9.1 Types of game data scientists. (Rodriguez, 2020.)

best approach, which creates a terminal revenue model based on the game rules and level map. The terminal revenue model, on the other hand, is no longer applicable when the players become multiplayers. Based on the terminal revenue model, a multiplayer model was created, which used a cooperative game theory as quantified in this way into practical games. Game theory has exceptional impacts on the theory of economics, and applying game-theoretical and relevant approaches to economics is growing very fast (Nie et al., 2014).

In modern ages, the foreign exchange (FOREX) business has brought quite a lot of analysis from researchers all over the world. Due to its weak features, different types of research have been conducted to achieve the task of predicting impending FOREX currency prices correctly (Islam et al., 2020). A huge amount of dollars is patronized every day on the foreign exchange (forex) business, making it the most financial business in the world. Precise forecasting of forex rates is a fundamental element in any country's strategy in the forex market (Galeshchuk & Mukherjee, 2017). Mohapatra et al. (2019) considered three different types of time series such as exchange rates, stock indices, and net asset values to predict forex market based on distributed incremental and diffusion-based learning strategies. A FOREX trading model focused on moving average forecast aggregation and metaheuristic optimization is presented in Papatsimpas et al. (2020). Three moving average forecasters are used in the model, which are optimized using the particle swarm optimization algorithm. It also comes with a collection of technical analysis-based trading guidelines. For the EUR/USD currency pair, simulation results based on real-world data are presented. Lulu and Jing (2019) focused on the supply and demand of private financial imbalances, as well as the nonmacro effect of asymmetric information on government regulation, in order to better understand the object, subject, and duty rights of private finance. The private financial subject and object loan model is developed using game theory, and the game is constructed based on the possible constraints between the subject and object in the demand financial relationship, as well as possible outcomes.

The instability of the Naira against foreign currencies has been a major setback sabotaging the revival process of stabilizing the country's economy. This is coming because of the high importation of goods that necessitated having an exchange of Naira with other foreign currencies, particularly the Dollar. Nigerian economy solely relies on the exportation of crude oil. However, Nigeria imports refined petroleum products for internal consumption, which requires the use of billions of dollars. Moreover, there is no strike balance between what imports from and exports to foreign countries. The stabilization of Nigeria will help strengthen the country's economy by making the right policy decision. Such a decision encompasses the domestic refining of crude oil and exploration of other means such as the Industrial Revolution, Agricultural Sector, and Information Technology & Computing, among others. Because of fluctuation of Naira, foreign investors are scared of coming to invest. Moreover, even some citizens rather invest outside Nigeria than investing locally. This proposed model is different from most of the research studies as they focused more on the prediction of the foreign exchange rate. While this is first resolved, the conflicting strategies use theory of games and then use machine learning model for prediction/forecasting.

9.3 DATA UNDERSTANDING/PROBLEM STATEMENT

Figure 9.2 shows the trend in which the Central Bank of Nigeria (CBN) projects a huge amount of dollars to stabilize Naira through its subsidy policy so that importers can source out dollars at a cheaper rate. A significant part of the crude oil is exported for refinement while importing the refined oil for internal consumption. The importation of refined oil has a tremendous impact on how Nigeria demands foreign currencies, particularly Dollar. Figure 9.3 shows the trends of crude oil export and production. The data show that almost 80% of the oil production is exported for refinement.

The major source from which Nigeria gets dollars is through the export of crude oil as shown in Figure 9.4. Nevertheless, this crude oil is not stable as the highest it reached is for the year 2008 where the price was $138.7/b (per barrel) but then drastically falling down and gradually growing up. In the early 2020, the price of crude oil was $14.3/b. The consequences behind insufficient dollars will be that the CBN cannot

Imbalance Stabilization between the Naira and the Dollar

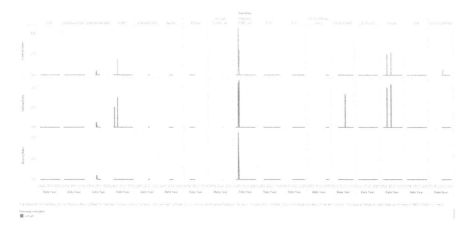

FIGURE 9.2 Trends of buying, selling, and CBN rate for foreign currencies.

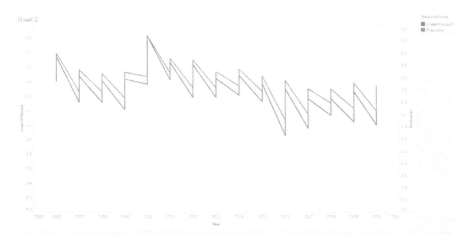

FIGURE 9.3 Trends of crude oil export and production for year.

give international businesspeople enough dollars to import most of what Nigeria can or not produce enough that can cater to the needs of its citizen which as such resulted in high inflation. The stability of Naira will bring balance with foreign currencies.

This paper tends to use a concept of game theory to address the issue of the instability of Naira against the Dollar with the help of machine learning techniques. The two currencies are conflicting in our day-to-day businesses. This will give us an idea of what we need to do to bring stability to Naira by solely reducing the dependency on a Dollar. The Governor of Central Bank of Nigeria (CBN), Godwin Emefiele, has revealed that the CBN has had the option to continue the strength of Naira due to the $60 billion inflow from the Investors' and Exporters' Foreign Exchange (I&E Forex) Window (Olalekan, 2019). This research will also give more insight into what is the optimum amount needed to project to strengthen or stabilize Naira against the Dollar using game theory and machine learning techniques.

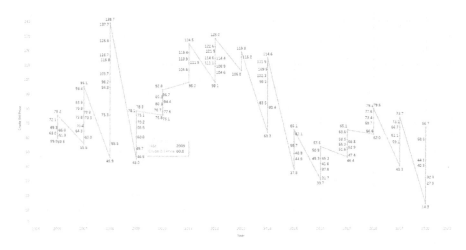

FIGURE 9.4 Trends of crude oil price for year.

9.4 CONCEPTUAL FRAMEWORK

This tends to use the concept of maxi-min and mini-max principles of game theory. Suppose the letter D represents a Dollar (USD), which is assumed a winner and N represents Naira (NGN), which assumed a looser. A game theory problem can be conceptualized and represented in a matrix form containing rows and columns as shown in Table 9.1 below. If D chooses it first strategy, it has a worse outcome of -10, if it chooses the second strategy, its worse outcome is -7 and lastly, if it chooses the third strategy, it has a minimum gain of $+5$. Among the three strategies of D, the third one tends to favor it more but only if N chooses the first strategy. However, there is no guarantee that N will choose the first strategy to favor D. N may choose the second strategy that has a better outcome, which will serve as a detriment to D. Similarly, N has a minimum loss if it chooses the first strategy but has maximum losses if it chooses the second strategy. Hence, N chooses the first strategy, thereby making D have more gain while it minimally loses. As such, D will always choose the first strategy to assure its minimum gain, and N will always choose the first strategy to minimize the losses. Thus, D is regarded as maxi-min competitor, as

TABLE 9.1
Matrix Representation of Two Competitors

		N	
		I	II
D	I	−10	−15
	II	−7	8
	III	5	12

it maximizes its minimum gains. In addition, N is said to maximize its maximum loses, also known as the mini-max competitor.

Mathematically, suppose $\left[d_{ij}\right]_{m\times n}$ denotes the payoff matrix of a game, then mini-max for D and maxi-min for N can be represented as $\text{Max}_i\left[\min_j d_{ij}\right]$ and $\text{Min}_j\left[\max_i d_{ij}\right]$, respectively.

Let $Max_i\left[\min_j d_{ij}\right] = d_{ab}$ and $Min_j\left[\max_i d_{ij}\right] = d_{tv}$, then d_{ab} is the minimum value in the ath row; therefore, $d_{ab} \leq d_{av}$ (another value in the tth row). Similarly, d_{tv} is the maximum value in the tth column. Hence, $d_{av} \leq d_{tv}$. This can be combined to get

$$d_{av} \leq d_{tv} = \text{Max}_i\left[\min_j d_{ij}\right] \leq \text{Min}_j\left[\max_i d_{ij}\right]$$

The maxi-min of d_{ij} is said to be the lower value of the game and is represented as $\underline{p}$, and the mini-max of d_{ij} is called the upper value of the game represented as $\overline{p}$. Moreover, the value of the game (p) is always between $\underline{p}$ and $\overline{p}$, which satisfies the inequality $\text{Max}_i \min_j$ for $D \leq \text{Min}_j \text{Max}_i$ for N or simply as $\underline{p} \leq p \leq \overline{p}$.

Finally, the intersection between the row and column where the row minimum is equal to the column maximum is called the saddle point of the game, which represents the value of the game.

9.5 CONCEPTUAL MODEL

Economy policy also known as stabilization measure is a collection of measures injected to study a financial system or economy. This policy direction symbolizes the stability measures, which are the fiscal policy and monetary policy (Anochie & Duru, 2015). Economic policies encompass choices made by government about spending and taxation, redistribution of earnings from wealthy to poor, and the inflow and outflow of money. According to Pettinger (2018), economy policy is classified into monetary policy, fiscal policy, microeconomic policies (tax, subsidies, price controls, housing market, regulation of monopolies), tariff/trade policies, supply-side policies, and labor market policies. Any time there is a state of affairs with two or more players that includes regarded pay-outs or quantifiable results, the game theory applies to assist in determining the maximum possible results. Figure 9.5 is a conceptual model that describes how game theory and unsupervised machine learning can be used to help the government in decision-making policy in improving the economy and consequently to strengthen Naira against Dollar. A decision is made by setting some policies as aforementioned to guide how an economy can be improved. The concept of game theory will be used to resolve conflict in the policies set in place. The decision made which is guided by the resolution will be used in real time to train and build a model to predict the GDP.

9.6 CONCLUSION

The stability of Naira against Dollar has been a major setback that affects the Nigerian economy. This is coming due to the high demand for dollars to import many things that Nigeria does not provide or manufacture domestically. This research tends to

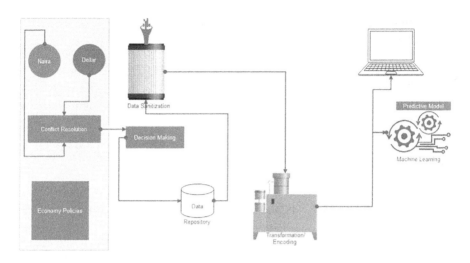

FIGURE 9.5 Conceptual model.

suggest the best precautionary measures to curtail the sudden devaluation of Naira using game theory and machine learning techniques. The outcome of this research will help the government in making the right decision. Moreover, having stable exchange rate will attract foreign investors and encourage local investors. In the future, each sector of the economy can be analyzed to discover the hidden contribution toward stability of exchange rate and consequently improve the GDP so that a priority can be given to that and lastly, right decision-making policy can be taken.

REFERENCES

Agrawal, A., & Jaiswal, D. (2012). When Machine Learning Meets AI and Game Theory. *IEEE Journal*, 5. http://cs229.stanford.edu/proj2012/AgrawalJaiswal-WhenMachineLearningMeetsAIandGameTheory.pdf.

Anochie, U. C., & Duru, E. E. (2015). Stabilization Measures and Management of the Economy: The Case of Nigeria. *International Journal of Development and Economic Sustainability*, 3(3), 1–12. http://www.eajournals.org/wp-content/uploads/Stabilization-Measures-And-Management-Of-The-Economy-The-Case-Of-Nigeria.pdf.

Aurélien, G. (2019). *Hands-on Machine Learning with Scikit-Learn, Keras, and TensorFlow* (2nd ed.). O'Reilly Media, Inc., Sebastopol, CA.

CFI. (2020). *Foreign Exchange—Definition, Trading Factors, Forex Markets*. Corporate Finance Institute. https://corporatefinanceinstitute.com/resources/knowledge/finance/foreign-exchange/.

Damer, S., Gini, M., & Rosenschein, J. S. (2019). The Gift Exchange Game: Managing Opponent Actions (Vol. 3). *AAMAS '19: International Conference on Autonomous Agents and Multiagent Systems*, Montreal, QC, Canada.

Economist. (2017, February 2). Nigeria's Foreign-Currency Shortage. *The Economist*. https://www.economist.com/middle-east-and-africa/2017/02/02/nigerias-foreign-currency-shortage.

FMDQ Securities Exchange. (2020). *Foreign Exchange*. FMDQ Group. https://fmdqgroup.com/markets/products/foreign-exchange/.

Galeshchuk, S., & Mukherjee, S. (2017). Deep Networks for Predicting Direction of Change in Foreign Exchange Rates. *Intelligent Systems in Accounting, Finance and Management*, 24(4), 100–110. doi:10.1002/isaf.1404.

Ippolito, P. P. (2019, September 20). Game Theory in Artificial Intelligence. *Medium*. https://towardsdatascience.com/game-theory-in-artificial-intelligence-57a7937e1b88.

Islam, Md. S., Hossain, E., Rahman, A., Hossain, M. S., & Andersson, K. (2020). A Review on Recent Advancements in FOREX Currency Prediction. *Algorithms*, 13(8), 186. doi:10.3390/a13080186.

Isma'il, M., Haruna, U., Aliyu, G., Abdulmumin, I., & Adamu, S. (2020). An Autonomous Courses Recommender System for Undergraduate Using Machine Learning Techniques. *2020 International Conference in Mathematics, Computer Engineering and Computer Science (ICMCECS)*, 1–6. doi:10.1109/ICMCECS47690.2020.240882.

Lulu, H., & Jing, G. (2019). Research on Private Financial Supervision from the Perspective of Game Theory. *Proceedings of the 2019 Annual Meeting on Management Engineering*, 96–106. doi:10.1145/3377672.3378042.

Ma, J. (2020). Optimal strategy solution based on Cooperative Game Theory. *Proceedings of the 2020 4th International Conference on Electronic Information Technology and Computer Engineering*, 1075–1079. doi:10.1145/3443467.3443907.

Mohapatra, U. M., Majhi, B., & Satapathy, S. C. (2019). Financial Time Series Prediction Using Distributed Machine-Learning Techniques. *Neural Computing and Applications*, 31(8), 3369–3384. doi:10.1007/s00521-017-3283-2.

Nie, P., Matsuhisa, T., Wang, X. H., & Zhang, P. (2014, May 19). Game Theory and Applications in Economics. *Journal of Applied Mathematics*. doi:10.1155/2014/936192.

Olalekan, F. (2019, December 2). CBN Governor Addresses Stability of Naira. *Nairametrics*. https://nairametrics.com/2019/12/02/emefiele-addresses-stable-naira/.

Papatsimpas, M. G., Lykogiorgos, I., & Parsopoulos, K. E. (2020). FOREX Trading Model Based on Forecast Aggregation and Metaheuristic Optimization. *11th Hellenic Conference on Artificial Intelligence*, 215–223. doi.:10.1145/3411408.3411415.

Pettinger, T. (2018). Different Types of Economic Policies. *Economics Help*. https://www.economicshelp.org/blog/1011/recession/solving-current-economic-problems/.

Rama, M. P. (2007). *Operations Research* (B. N. V. Jagdeesh, Ed.; 2nd ed.). New Age International (P) Ltd, New Delhi.

Rekha, J. U., Chatrapati, K. S., & Babu, A. V. (2016). Game Theory and Its Applications in Machine Learning. *Information Systems Design and Intelligent Applications*, 435, 195–207. doi:10.1007/978-81-322-2757-1_21.

Rodriguez, J. (2020, February 27). A Crash Course in Game Theory for Machine Learning: Classic and New Ideas. *Medium*. https://towardsdatascience.com/a-crash-course-in-game-theory-for-machine-learning-classic-and-new-ideas-50e33ba2636d.

Routledge. & S. Christian. (2004). *Game Theory and Economic Analysis*. Taylor & Francis e-Library, London.

Serdar, G. (2012, June 21). A Short Note on the Use of Game Theory in Analyses of International Relations. *E-International Relations*. https://www.e-ir.info/2012/06/21/a-short-note-on-the-use-of-game-theory-in-analyses-of-international-relations/.

Sidehabi, S. W., & Tandungan, S. (2016). Statistical and Machine Learning Approach in Forex Prediction Based on Empirical Data. *2016 International Conference on Computational Intelligence and Cybernetics*, 63–68. doi.:10.1109/CyberneticsCom.2016.7892568.

Stefano, A. D., Maesa, D. D. F., Das, S. K., & Liò, P. (2020). Resolution of Blockchain Conflicts through Heuristics-Based Game Theory and Multilayer Network Modeling, 10. *ICDCN 2020: 21st International Conference on Distributed Computing and Networking*, Kolkata.

Thu, T. N. T., & Xuan, V. D. (2018). Using Support Vector Machine in FoRex Predicting. *2018 IEEE International Conference on Innovative Research and Development (ICIRD)*, 1–5. doi:10.1109/ICIRD.2018.8376303.

Usha, B. A., Manjunath, T. N., & Mudunuri, T. (2019). Commodity and Forex Trade Automation Using Deep Reinforcement Learning. *2019 1st International Conference on Advanced Technologies in Intelligent Control, Environment, Computing & Communication Engineering (ICATIECE)*, 27–31. doi:10.1109/ICATIECE45860.2019.9063807.

10 The Emerging Role of Big Data in Financial Services

Deepika Dhingra and Shruti Ashok
Bennett University

CONTENTS

10.1	Introduction	175
10.2	Literature Review	176
10.3	Research Methodology	178
10.4	What Is Big Data?	178
	10.4.1 Evolution of Big Data	178
	10.4.2 Big Data Lifecycle	179
10.5	Reasons for the Proliferation of Big Data	179
10.6	Big Data Application: Need in Financial Industry	180
10.7	Big Data: Applications in Financial Services Sector	181
10.8	Application of Big Data for Crises Redressal in Banking	183
10.9	Big Data Analytics Application – Used Case from Indian Banking	185
	10.9.1 HDFC Bank – Analytics to Provide a Comprehensive Understanding of Customers	185
	10.9.2 ICICI Bank – Use of BI and Analytics to Reduce Credit Losses	185
	10.9.3 Axis Bank – Analytics for Customer Intelligence and Risk Management	186
	10.9.4 State Bank of India – Using Data Analytics	186
	10.9.5 ING Vysya Bank – Need for BI Implementation	186
10.10	Big Data Constraints in Financial Services Sectors	186
10.11	Conclusions	187
10.12	Scope of Future Research	188
References		188
Web Reference		189

10.1 INTRODUCTION

In this era of technology and innovation, data is perceived as one of the most valued commodities for businesses. Every human activity over the last few decades, has witnessed advancement in technology and financial markets. Continuous innovations in financial sectors has enabled Big Data technology to emerge as an essential and fundamental part of the financial service industry. Applications of Big Data cover varied financial services such as online credit facility, SME finance, crowd-funding platforms, cryptocurrency, asset management, wealth management platforms,

DOI: 10.1201/9781003140474-10

mobile payments platforms, etc. All these service transactions generate huge amount of data every single day and any loss or mutilation of this data can result in serious complications for that vector of the financial services industry. The efficient application of available Big Data helps industries to decide the financial products that they should propose.

The market for big data technology in the financial and insurance domains is one of the most promising. This chapter examines the transformation that Big Data analytics has brought not only for individual business operations but for the entire finance sector. It is expected that the current study will address diverse financial outlooks and provide Valuable insights on the relevance of big data to the readers.

Existing research on the impact of Big Data in financial services is not as broad as research on other financial areas. Current studies have primarily focused on limited topics of Big Data without elaborating on its impact and prospects in the area of finance. Research exploring the interdependence of Big Data and financial services is extremely fresh; hence, limited studies have focused on the applications of Big Data in varied financial research contexts. This study, therefore, attempts to identify areas where applications of Big Data can have a potential influence on the decision-making process, both from the point of view of institutions as well as customers. This study is also expected to link diverse perspectives on Big Data and financial events as provided by existing researchers and academicians.

After having discussed the relevance of the present study, the remainder of this chapter is organized as follows: Section 10.2 provides a brief review of the existing studies conducted on the role and need of Big Data in financial services management. Section 10.3 explains the research methodology adopted. Section 10.4 discusses evolution of Big Data and its lifecycle. Section 10.5 details the reasons for the increase in the quantum of Big Data, while Sections 10.6 and 10.7 elaborate on the need and applications of Big Data in the financial industry, respectively. Section 10.8 explains the applications of Big Data in redressal management at banks, and Section 10.9 discusses cases of Big Data application in six Indian banks. Section 10.10 elaborates on the constraints of Big Data followed by conclusion in Section 10.11. The chapter concludes with suggestions for future research as the last section.

10.2 LITERATURE REVIEW

Big Data has always had a considerable impact on different functions of businesses like management of human resources, business process, research and development, business analytics, and other business fields. Rabhi et al. (2019) positioned Big Data as an important element in the management of business processes that facilitates decision-making. Their study also mentions the three analytics approaches, i.e., prescriptive analytics, descriptive analytics, and predictive analytics that makes the conventional process of data analytics more robust. Duan and Xiong (2015), Grover and Kar (2017), Ji et al. (2019), and Pappas et al. (2018) deliberated on importance of Big Data in business analytics. All of these studies identified Big Data as an aid that overcomes various business challenges and enhances data management utilizing system infrastructure, encompassing techniques to capture, accumulate, transmit, and process data. Choi and Lambert (2017) found "Big Data" to be more relevant and significant for risk analysis

Duan and Xiong (2015) discovered that top-performing companies take their business decision by applying analytics rather than intuition. They also concluded that business analytics need to be intertwined with business strategies to Provide enhanced analytics-driven insights. Grover and Kar (2017), cited the example of companies such as Facebook, Apple, Amazon, Google, and eBay, discovered that these organizations utilize digitized transaction data like time of the transaction, quantity of purchase, price of the product, and other customer credentials and information consistently to assess the condition of the market to enhance business operations. Holland et al. (2019) demonstrated the realistic and theoretical influences of Big Data in businesses like establishing B2B relationships by accounting for consumer search patterns. Big Data analytics was found to be not only considerably enhancing sales growth (measuring through monetary outcomes) but also improving customer relationship performance (nonmonetary outcomes).

Belhadi et al. (2019) through a case study analysis of two firms observed that through application of Big Data analytics, the company considered for analysis aimed for a qualitative jump, which enabled its teams to develop and improvise prototypes of its turnkey factories in Africa. This study focused on the increasing applications of Big Data analytics in manufacturing process, aiding enhanced performance that accelerates-decision-making. Cui et al. (2020) in their study stated the four most common Big Data applications that are critical to realizing smart manufacturing processes, i.e., monitoring, prediction, ICT framework, and data analytics. Shamim et al. (2019) mentioned the emergence of Big Data as an opportunity to enhance a firm's performance.

Yadegaridehkordi et al. (2020) discussed the positive impact of Big Data on a company's productivity, citing the benefits of its adoption for policymakers, governments, and businesses by getting useful insights and making informed decisions.

Raman et al. (2018) integrated supply chain management with Big Data and devised a new model, Supply Chain Operations Reference (SCOR). The model revealed the utility of this technology in enhancing value and creating gains for the industry. This model not only evaluates the supply chain's financial performance but also provides a system that enhances practical decision-making for the firm. Lamba and Singh (2017) focused on factors responsible for decision-making of the supply chain process and specified that decision-making, based on data, helps to attain better results in the management of logistics activities, enhancement in process, cost optimization, and improved management of inventory. Sahal et al. (2020) and Xu and Duan (2019) showcased the connection between stream processing platforms and cyber-physical systems for Industry 4.0. They found that Big Data supports and facilitates attainment of the most essential goals of Industry 4.0 applications – enhancing the production efficiency while minimizing the cost of production.

Duan and Xiong (2015) concluded that compared to structured data, Big Data analytics have a greater significance in comprehending unstructured data like time-series data, graphs, and text both for data storage and data analytics methods. Zhao et al. (2014) acknowledged two main challenges of Big Data analytics – assimilation of internal and external data – linking datasets across data sources and identifying the most appropriate data for analysis. Hofmann (2017) discussed the challenges of storing and processing various types of data and hence stressed the need and significance of Big Data analytics.

Review of all the above-mentioned studies on Big Data brings forth the point that most of the research conducted has focused on the application of Big Data in business decision enhancement. Very few studies have explored the various applications of Big Data in finance or the reliance of finance on Big Data analytics.

10.3 RESEARCH METHODOLOGY

With limited available literature on Big Data and finance, this study focuses on identifying the immense scope and application of Big Data in the financial world. The present study employs a qualitative approach to analyze the impact of Big Data on financial services and the strategies adopted by existing as well as new players. To accomplish this, secondary data sources were used to collect related data. The study uses electronic databases like Scopus, the web of science, Google Scholar, and some unpublished research articles that were deemed relevant for the study.

10.4 WHAT IS BIG DATA?

"Big Data" is one of the most common buzzwords used in the financial services industry today, often referring it as a substitute for customer/real-time/predictive analytics.

"Big Data" can be defined as a collective term used for present-day methodologies and technologies that engage in collection, organization, processing, and analysis of huge, assorted (structured and unstructured) and multifaceted sets of data, while the different facets of analytics refer to the investigation conducted on these data sets to identify patterns and generate value for businesses.

Big Data is characterized by three Vs, i.e.,

- **Volume**: Handling of huge amount of data in terabytes or petabytes. Such huge data cannot be processed with conventional data processing tools within short and precise time limits.
- **Velocity**: The real-time data and swift and accurate analysis for insight can be essential for businesses. Therefore, Big Data technologies should be capable of processing the data in both batch and real time.
- **Variety**: Big Data technologies should support various kinds of data, i.e., ranging from highly structured data to unstructured information such as audio, video, text, tweets, blogs, and Facebook status.

10.4.1 EVOLUTION OF BIG DATA

Big Data is believed to be an enthralling and captivating topic and is evolving in almost every sector of the business world. Its evolvement in finance is evident with its impact on areas such as risk management, financial management, financial analysis, and financial data applications management.

Every organization collects huge data in billions of pieces every day. The need for Big Data techniques to deliver faster, impartial, and balanced estimates becomes pertinent to handle this massive amount of database. Big Data can help firms assess their risk, which is one of the most important aspects affecting profitability. The evolution of Big Data has extended the range of data types that can be managed, empowering financial institutions and banks to espouse and react in a more sophisticated manner both physically and through digital interactions with their existing and prospective customers.

10.4.2 BIG DATA LIFECYCLE

Figure 10.1 shows that Big Data follows a lifecycle pattern. Starting with the Generation of Data and ending at Visualization of Data, the process flows through Data Acquisition, Data Storage, and Data Analytics.

10.5 REASONS FOR THE PROLIFERATION OF BIG DATA

Evolution of technology has led to data input in larger amounts, and the need to process huge quantity of diverse and complex data in real-time has increased manifold. As data sets are emerging larger and more complexed, conventional tools are no more capable of processing this data adequately and at lower cost and within realistic time. The following technological advancements have contributed majorly to the proliferation of Big Data:

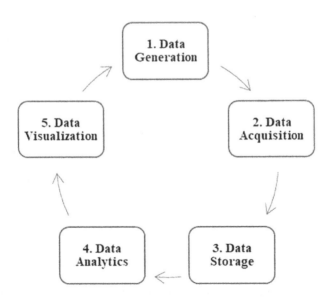

FIGURE 10.1 Big Data life cycle. (Rabhi et al., 2019.)

- The upsurge of IoT (Internet of Things) has blown up quantity of data, consequently adding new, continuous (even if customers are not engaging with financial institutions) data streams.
- Advancement in authentication techniques, like biometric and continuous authentication, have resulted in considerable upsurge in real-time data.
- The growth in Open Architectures (Open APIs) platforms has allowed financial organizations to procure valued data of their customers from competitor's data storage.
- The increased competition and recent success of Fintech companies servicing their customers using Big Data technologies demonstrate that Big Data is enhancing customer service.
- As the financial fraud and crime graph is soaring, financial organizations need to safeguard the "trust" of the customers that they have on the institutions. That further intensifies the pressure on financial institutions to protect customer data through various security technologies.

10.6 BIG DATA APPLICATION: NEED IN FINANCIAL INDUSTRY

Financial services companies have been incessantly aiming to integrate Big Data tools to recognize concealed customer sentiments on real-time basis. By using Big Data techniques, financial institutions can enjoy the advantage of enhanced and precise credit risk assessment. Big Data applications assist to measure the credit risks of banks in home equity loans. It includes correlating internal and external data that can aid in fraud detection by minimizing manual intervention. Improved data handling allows risk analysts to develop effective and market risk specific plans to manage risks better. Big Data is very useful for banks to adhere to and act in accordance with regulatory and legal guidelines in the domains of integrity and credit risk. It helps financial organizations/lenders to decrease risks by forecasting a client's capability to repay their loan. This helps people to get greater access to the credit facility, enabling banks to minimize their credit risks.

The following sections discuss the need for Big Data applications in the financial industry.

Access to Structured and Unstructured Data: Big Data technologies offer an inclusive view of the ecosystem by providing augmented, integrated, and structured transactional data and provide a comprehensive view of customer behavior and psychography. These tools can empower organizations to assess greater amounts of both structured and unstructured data within a shorter time frame. Any kind of data, whether structured, semi-structured, or unstructured data like emails and SMS messages can be examined to discover occasional customer sentiments and behavioral patterns. Financial services companies can deploy Big Data tools and apply the results to recognize customer segmentation to assist product development and enhance customer service.

Multiple relationships with bank and customer insight: In the present scenario, customers maintain relationships with various financial organizations and banks. Therefore, a particular or single bank does not have clear understanding about customer's behavior and its buying and spending patterns. Having said that, it is critical

Emerging Role of Big Data 181

to get an all-inclusive picture of the customer's behavior to satisfy and retain them. Big Data analysis seems to be a possible solution wherein banks can acquire customer behavior data and information from all possible sources such as from call centers, customer emails, customer's social media posts and the insurance claims on a common platform.

Categorization of a maintained data: Almost all banks and financial services companies maintain their data in silos and under different heads like savings accounts, current account, term deposits, term loans, housing loan, car loan, etc. From fraud detection, credit monitoring, and mitigation to offering customers superior deals, having comprehensive customer information in a single thread is valuable and useful in all aspects. Even the calculations of loan default and assessment of risk is possible only with an all-inclusive single thread customer data for applying Big Data techniques. Using Big Data technologies, better rate of interest can also be considered and offered. To achieve a successful outcome for effective marketing, it is necessary to understand and anticipate customer's requirements precisely in real time to fulfill the same exactly at the right time.

Exponential data growth: Increase in volume of financial transactions has led to data surge in financial companies. This growth in the number of companies operating in the financial space has accelerated after the propagation of electronic/online trading in capital market transactions.

Increasing scrutiny from regulators: With better implementation of regulatory mechanisms, it is becoming imperative to achieve better transparency in financial transactions. Hence, financial establishments require accurate analyzis of raw data to meet the regulatory terms and requirements.

Advancements in technology leading to increased activity: Digitization of financial products has eased and made online financial transactions affordable. Now people can trade through various kinds of accounts more often with the click of a button, resulting in creation of huge amounts of data.

Changing Business Models: Motivated by numerous elements, markets today are fundamentally different from what they were few years back especially for financial organizations. Acceptance and implementation of Big Data analytics is required to form business models for financial organizations that help to retain market share in an an all-pervasive competitive environment.

10.7 BIG DATA: APPLICATIONS IN FINANCIAL SERVICES SECTOR

With customers today using multiple digital platforms for their online transactions, trend analysis based on historical data is becoming obsolete and is being replaced by comprehensive real-time data analysis encompassing all forms of customer management and relationship pricing. Being a data driven industry primarily, it is expected that Big Data and Customer Analytics can add immense value to the financial sector.

The key Big Data analytics applications in banking and financial services sector are:

- **Improved risk management**
  By providing real-time and better insights of customer behavior, big data can help financial services providers enhance risk management substantially.

- **Detection & Prevention of cyber (Identity) frauds**
  For continuous identification of risks and assessment related possible frauds and adequate safety measures, financial companies can use Big Data analytics to provide inputs to fraud detection engines.
- **Liquidity risk management**
  Big Data analytics facilitates banks to obtain insights for management of liquidity. These insights can be valuable for banks and other financial companies for improved liquidity management.
- **Credit risk management**
  Customer insights provided by Big Data analytics aid in improving credit models for corporate and private customers, enhancing their credit score. These insights can be procured from historical transactions, public information (e.g. company's annual reports), and IoT data (e.g. car/home/inventory sensors).
- **Cards fraud detection**
  Big Data can analyze the patterns of transaction (timing, location, amount, and merchant type) for identification of fraudulent transactions, so that they can be blocked.
- **Insurance fraud detection**
  At the time of signing up a new policy or at the time of claims, Big Data analytics facilitate improved fraud detection and prevention. Various pointers like IoT data (sensor can directly inform regarding home damage or car accident), numerous simulation attempts by customers for an insurance policy, and unpaid premiums can be used to detect any maligned interest toward insurance claim.
- **Managing Legal Claim**
  Big Data can assist in formulating and responding to legal issues involving huge data, in a more systematic and lucid manner, e.g., information regarding legal matter, case assessment, and enhancing tracking guidelines and procedures to evade fines and sanctions.
- **Personal Financial Management**
  Big Data can be used for automatic classification of financial transactions in different groups. This can enable customer segmentation, recommendation of suitable budget plans in accordance with customer's objectives.
- **Sentiment and Predictive Analysis using Big Data**
  Sentiment and predictive analysis emanating from Big Data can empower banks and financial service companies to focus on the challenges of customer retention. These technology-based tools allow the companies to offer tailor made financial products suiting customer's needs and build strategies according to customer expectations, empowering companies to detect fraud patterns enabling reduction of credit risk.
- **Personalized Advise on Wealth Management**
  Big Data analysis can be applied to understand customer's objectives and financial goals, family circumstances, risk averseness, financial conditions, and propose financial planning, investment advice, and tax advice, based on insights gathered.

- **Customer feedback Consolidation**

  Applying the tools of sentiment analysis can empower financial companies and banks to detect ground-breaking propositions to enhance the products and envisage behavior of the customer. By analyzing customer's sentiments data obtained from social media platforms and interactions, banks can get access to customer's feedback about their service offerings. Assessing data on real-time basis, sentiment analysis allows banks to take quicker decisions to manage customer's negative feedback or sentiments that might have appeared on social networks. Effective and real-time monitoring and capturing of this data can help banks evaluate the probable and hidden effect of their decisions. Big Data enables banks in linking the customer feedback with their communications that reflects their emotions, providing key inputs to aid strategic decision-making.

- **Implementing Loyalty and Reward Programs**

  With the evolution of Big Data tools and techniques, identification of customer's likes, dislikes, and preferences is possible. Sentiment analysis plays a major role in the background of reward and loyalty programs for customer retention and product/service attractiveness. Thorough examination of the customer confidence, provided by specific data elements, financial companies and banks can gauge the customer's frame of mind and preferences and can reward loyal customer's accordingly. This information contributes significantly to financial product and service improvements, enabling the bank/financial institution to achieve a competitive advantage over its rivals in the market.

- **Capitalizing on Customer Insights**

  Predictive analytic techniques using Big Data can enable the users to source huge historical data to envisage the events that are likely to take place in the future. By enquiring, envisioning, and reporting these historical datasets, financial companies and banks can get substantial understanding of instructive, transactional and behavior patterns of customers. Such insights can help these companies/banks to take strategic decisions on the prospective products/services that can be offered to the customers.

- **Model Building**

  Using customer spending behavior and product usage data, Big Data can empower banks/financial institutions to create models. These models help identify the most popular products/services among the customers and enables the bank/financial institution to focus on delivering them more effectively. Model building can thus aid banks to augment their share of income, improve customer loyalty, also intensify their profitability.

10.8 APPLICATION OF BIG DATA FOR CRISIS REDRESSAL IN BANKING

Incase a bank/financial company is facing customer migration due to lack of customer satisfaction, Big Data analytics can play a noteworthy role in crisis redressal in the following ways:

- Identify the reason for the fall in customer satisfaction.
- Analyze the cardholder's spending patterns.
- Analyze the channel usage of the customer—credit/debit/ATM cards.
- Study the customer's behavior and promote cross-sell based on customer's needs.

The crisis redressal process starts with examining the measurement data of customer satisfaction to recognize the most plausible cause of fall, whether because of poor services or any other issues. Using classification techniques, the relevant customer segment is identified, and analysis of feedback is considered to detect the reasons behind the fall in satisfaction. These results help find the most suitable financial products that can then be recommended to the distraught customer based on their segment.

- **Feedback Analysis**

  In any organization, feedback analysis plays a pivotal role in problem identification and in finding the most appropriate solution. In bank/financial services firms, feedback can be taken from customers visiting the bank-both physically and virtually. The customers can be asked to rate the bank's services on the following parameters:
  - Quality of service
  - Service speed
  - Response to their queries
  - Branch availability
  - System reliability.

  Data collected on the parameters mentioned above can be analyzed by plotting a graph, and this analysis can throw light on the steps that can be taken by the bank to improve its services.

- **Online Transactional Analysis**

  Analyzing the online transactions of customers can provide valuable insights into their spending patterns. Depending on the spending patterns, cause can be identified that can lead to suggestions or recommendations to improve customer experience.

- **Channel Usage Analysis**

  Channel usage patterns based on expenditure channel (online transactions versus ATM or card) can provide insights into customer behavior. For instance, if certain customers are found having surplus funds, investment plans can be offered to them accordingly.

- **Analysis of Consumption and Expenditure Patterns**

  Depending on a customer's consumption/spending patterns, specific banking products can be cross sold to the customer. For example: In case a customer is found to have high spending appetite, s/he can be offered a credit card or a scheme.

- **Security and Fraud Analysis**

  Probable threats to banks/financial institutions can be recognized based on the customer's historical patterns of transactions and consumption. Frauds that have been committed in the past can be easily identified through Big Data analytics and banks can be laced to take proactive steps accordingly, improving active and passive security of the bank.

10.9 BIG DATA ANALYTICS APPLICATION – USED CASE FROM INDIAN BANKING

With a view to derive maximum benefits, many international banks have deployed of Big Data analytics techniques across the areas of their operations, from financial crime management, regulatory compliance management, sentiment analysis, reputation risk management, product cross-selling, etc. With the fast-growing banking sector in India, the sector is also endeavoring to catch up with this development. An IBEF report mentions that by 2020, banking in India is poised to become the fifth largest banking industry in the world, and by 2025, it will be the third largest. In India, banks are using data analytics to attract new customers, customer retention, detecting opportunities for sales, and minimizing the losses. That is helping them to stay in competition.

10.9.1 HDFC Bank – Analytics to Provide a Comprehensive Understanding of Customers

According to a Live Mint article, one of the first cases using analytics was observed in the early 2000s when HDFC Bank Ltd started investing in technology and set up a data warehouse that would help in understanding enormous amorphous data netted by its information technology (IT) systems. Analytics engines help the banks to notice every typical financial trait of customer behavior, for instance, if the customer holds an active account or just a salary account. Analytical tools also give insights on customer's personal behaviors, which helps bank to offer suitable product and service accordingly. Data analytics reduce money laundering activities by identification of doubtful actions such as money rotation in multiple accounts, detecting large cash deposits in a single day, number of accounts opening within a short time, or any activity in accounts which were dormant for a long time. It also prepares the banks to notice customer's credit history before approving loans further.

10.9.2 ICICI Bank – Use of BI and Analytics to Reduce Credit Losses

The subprime mortgage disaster in the US (2007) had its consequences in India as well. Banks were required to deal with reduced liquidity challenges, increased rate of interest and at the same time ensure that customers remain loyal to the bank. ICICI bank acknowledged debt collection process as a key factor where a friendly approach could enhance the customer satisfaction. In the process of debt collection, for each case, appropriate customer-approach was opted. The management aimed to transform the process of debt collection as a tool to retain the customer and use the technology to accomplish the goal. In-house developed BI solutions deployed in

ICICI bank comprised of SAS components, TRIAD, Sybase, Posidex, Blaze Advisor, and Data Clean that factored in various parameters like efficiency of collector, risk behavior, exposure, and customer profile.

10.9.3 Axis Bank – Analytics for Customer Intelligence and Risk Management

Axis Bank has observed increased output of the sales force by five times in the last few years. The bank uses data analytics in almost every domain. For example, before visiting any prospective loan customer, the sales force does their homework analyzes the results provided by analytics and estimate the background of the customer and his loan appetite. The bank also uses technology to enhance customer loyalty and minimize loan prepayments. Using SAS, the bank provides customer intelligence throughout the institution. The SAS tool helps the bank to enhance risk management across the institution by sending early warning indications.

10.9.4 State Bank of India – Using Data Analytics

There is no doubt that banks from private sector are leading the change by using tools of data analytics for effectual decision-making, but at the same time, public sector banks are not far behind. Data warehouse of State Bank of India has over 120 TB of data and every day receives 4 TB of additional banking data. The bank uses data analytics for creating data models for automotive, education, SME, and housing loans. The modelling of the loans is intended to reduce the incidence of NPAs. An instance of application of Big Data analytics in loan modelling is use of historical data like information from past payment records, credit ratings agencies, income tax departments, CIBIL scores to identify the most appropriate borrower and remind them accordingly. SBI also actively applies data analytics to determine the most appropriate positioning of ATMs and new bank branches, to maximize efficiency and minimize costs.

10.9.5 ING Vysya Bank – Need for BI Implementation

When ING Vysya Bank (now taken over by Kotak Mahindra Bank) observed that various end - users attended meetings with imprecise reports, it felt the need of embracing business intelligence (BI). They needed a solution for users to generate precise and timely reports. They created a common data repository with the help of SAP BO, which facilitated users to get precise reports and enhance efficacy.

10.10 BIG DATA CONSTRAINTS IN FINANCIAL SERVICES SECTORS

The constraints for Big Data in the banking and financial sector can be summarized as follows:

Silos of Data: Existence of customer data in silos such as loan servicing, portfolio management, CRM, etc. pose significant legacy system challenges in data integration.

Skills and Development Challenges: Some organizations have recognized the opportunities presented byBig Data; however, shortage of skilled workforce continues to be a big impediment for them.

Successful implementation of Big Data technologies requires development of New skill sets: Data Specialist need to have good understanding of various programming tools like SAS/R/SQL/Python coupled with effective visualization skills. These employees need to be paid significantly higher salaries to select and retain them. Scouting for the right talent can be a challenging task for the banks.

Lack of Strategic Focus: Often, top management views Big Data as yet another IT project without giving due consideration to its merits. Considering the many benefits of Big Data applications, it is vital that the management recognizes the value that Big Data analytics can bring in, and plan for its successful implementation in the entire organization.

Privacy Concerns: Due to existence of sensitive and correlated customer information, there are concerns regarding privacy that limits its acceptance in financial institutions. Analysis of such sensitive and private data often becomes objectionable to the customers.

Change in Customer Behavior and Expectations: Customers are increasingly engaging digitally with the bank and other financial companies, which means personal contact has been decreased, but at the same time, it is possible to acquire much more customer data in an automated way (e.g., geo-location data, browsing history, time of the interactions). This data should be used to compensate decreased customer interaction, caused by the decline due to lack of personal engagement.

Absence of New Technology and Infrastructure: Many banks continue to be laced with outdated IT infrastructure containing data silos with many legacy systems. In such scenarios, implementing Big Data is looked upon as a mere add-on, rather than a new ingenuity. Many banks fail to recognize the efficacy of Big Data analytics and hence fail to appreciate its utility in improving the core business of the bank.

Data Quality: Big Data analytics has maximum impact on the Data Quality. Maintaining the quality of data processed through Big Data analytics is of paramount importance and hence poses a significant challenge to the banks. To ensure highest Data Quality and Integrity, banks should strictly ensure that all the data Quality characteristics—accuracy, validity, timeliness, completeness, and reasonableness—are well demarcated, measured, documented, and made accessible to end users. care should be taken to preserve the original data and that no data has been stolen or lost in the process.

Organizational Mindset: Banks in general need to develop a mindset with superior Data Curiosity, Data-Driven thinking coupled with the need to invest more in acquiring, storing, and analyzing data. It is a fact that the banks that take advantage of Big Data would stay ahead of its competitors. Thus, it becomes imperative that the management takes cognizance of its utility and give Big Data investment and implementation top priority.

10.11 CONCLUSIONS

Banks over the years have maintained massive customer data, but owing to certain limitations, this data has not been effectively utilized to provide valuable customer insights.

Considering severe competitiveness in the financial sector, if banks want to stay competitive, they need to implement a data-driven approach. As prospects for incumbent banks and financial institutions from these insights are limitless, Big Data will surely be a major differentiator in the future competitiveness of financial organizations.

Machine learning, Big Data, cloud computing, and AI are driving digitalization in the financial industry. Big organizations are acclimatizing to these new technologies to adapt a digital makeover, bolster profit, and meet customer demand. Although most organizations are storing new and valued data, the question that arises here pertains to the impact of this data in the financial services industry. In this perspective, all financial organizations are introducing innovative technologies and considering data as a prerequisite. Thus, basis the findings in this study, it is rational to conclude that Big Data has transformed the financial industry predominantly with real- time insights of stock market. It has changed the way of trading and investment. It has also brought major shift in detection and prevention of frauds and accurate analysis of risk by the machine learning process. These services are making an impact resulting in increased revenue and improved customer satisfaction, accelerating manual processes, improved purchase interface. Even with these innovative service conductions, there are some critical challenges of Big Data that are persistent in the world of finance. The issues of privacy and data protection are considered critical to data quality and in meeting the regulatory challenges of Big Data services.

10.12 SCOPE OF FUTURE RESEARCH

Every financial service is technologically inventive and considers data as its blood. Findings from this study conclude that even though big data has revolutionized the financial industry, providing real time insights, enabling fraud detection and precise risk analysis, yet it leaves abundant scope for future research. Discussions in this study outline the road for future research. Despite the fact that all financial products and services completely rely on data generation, yet limited research has been done on the application of Big Data in the field of finance. In times to come, considerable research will be essential to deal with technical issues like handling of large data sets. Future research should emphasize on creating easy access to large data sets for small firms. Attention should be given to explore Big Data's influence on financial markets, its products and services. Research in the field of security risk of Big Data is also essential in financial services. Particularly, research in the future should continue to explore Big Data's influence on stock market. To conclude, the evolving and upcoming challenges of Big Data in finance discussed in this study should be empirically highlighted in future research.

REFERENCES

Belhadi A, Zkik K, Cherrafi A, Yusof SM, El Fezazi S. Understanding big data analytics for manufacturing processes: insights from literature review and multiple case studies. *Comput Ind Eng*. 2019;137:106099. doi.:10.1016/j.cie.2019.106099.

Choi T, Lambert JH. Advances in risk analysis with big data. *Risk Anal*. 2017;37(8). doi:10.1111/risa.12859.

Cui Y, Kara S, Chan KC. Manufacturing big data ecosystem: a systematic literature review. *Robot Comput Integr Manuf.* 2020;62:101861. doi:10.1016/j.rcim.2019.101861.

Duan L, Xiong Y. Big data analytics and business analytics. *J Manag Anal.* 2015;2(1):1–21. doi:10.1080/23270012.2015.1020891.

Grover P, Kar AK. Big data analytics: a review on theoretical contributions and tools used in literature. *Global J Flex Sys Manag.* 2017;18(3):203–229. doi:10.1007/s40171-017-0159-3.

Hofmann E. Big data and supply chain decisions: the impact of volume, variety and velocity properties on the bullwhip effect. *Int J Prod Res.* 2017;55(17):5108–5126. doi:10.1080/00207543.2015.1061222.

Holland CP, Thornton SC, Naudé P. B2B analytics in the airline market: harnessing the power of consumer big data. *Ind Mark Manage.* 2019. doi:10.1016/j.indmarman.2019.11.002.

Ji W, Yin S, Wang L. A big data analytics based machining optimisation approach. *J Intell Manuf.* 2019;30(3):1483–1495. doi:10.1007/s10845-018-1440-9.

Lamba K, Singh SP. Big data in operations and supply chain management: current trends and future perspectives. *Prod Plan Control.* 2017;28(11–12):877–890. doi:10.1080/09537287.2017.1336787.

Pappas IO, Mikalef P, Giannakos MN, Krogstie J, Lekakos G. Big data and business analytics ecosystems: paving the way towards digital transformation and sustainable societies. *IseB.* 2018;16(3):479–491. doi:10.1007/s10257-018-0377-z.

Rabhi L, Falih N, Afraites A, Bouikhalene B. Big data approach and its applications in various fields: review. *Proc Comput Sci.* 2019;155(2018):599–605. doi:10.1016/j.procs.2019.08.084.

Raman S, Patwa N, Niranjan I, Ranjan U, Moorthy K, Mehta A. Impact of big data on supply chain management. *Int J Logist Res App.* 2018;21(6):579–596. doi:10.1080/13675567.2018.1459523.

Sahal R, Breslin JG, Ali MI. Big data and stream processing platforms for Industry 4.0 requirements mapping for a predictive maintenance use case. *J Manuf Sys.* 2020;54:138–151. doi:10.1016/j.jmsy.2019.11.004.

Shamim S, Zeng J, Shafi Choksy U, Shariq SM. Connecting big data management capabilities with employee ambidexterity in Chinese multinational enterprises through the mediation of big data value creation at the employee level. *Int Bus Rev.* 2019. doi:10.1016/j.ibusrev.2019.101604.

Xu L, Duan L. Big data for cyber physical systems in Industry 4.0: a survey. *Enterp Inf Syst.* 2019;13(2):148–169. doi:10.1080/17517575.2018.1442934.

Yadegaridehkordi E, Nilashi M, Shuib L, Nasir MH, Asadi M, Samad S, Awang NF. The impact of big data on firm performance in hotel industry. *Electron Commer Res Appl.* 2020;40:100921. doi:10.1016/j.elerap.2019.100921.

Zhao JL, Fan S, Hu D. Business challenges and research directions of management analytics in the big data era. *J Manag Anal.* 2014;1(3):169–174. doi:10.1080/23270012.2014.968643.

WEB REFERENCE

Bhasker G. Analytics in Indian Banking Sector – on a right track. Retrieved on 25th January 2021 from https://analyticsindiamag.com/analytics-in-indian-banking-sector-on-a-right track/#:~:text=ING%20Vysya%20(now%20acquired%20by,generate%20accurate%20and%20ti mely%20reports.

11 Digital Payments in India
Impact of Emerging Technologies

Manisha Sharma
NMIMS University

CONTENTS

11.1 Introduction 191
11.2 Digital Payments in India 192
11.3 Inhibitors of Digital Payments 193
    11.3.1 Social Risk 194
    11.3.2 Psychological Risk 194
    11.3.3 Time Risk 194
    11.3.4 Data Security Risk 194
    11.3.5 Overspending Risk 195
11.4 Facilitators of Digital Payments 195
    11.4.1 Easy and Convenient to Use 195
    11.4.2 Enabled Transaction from Anywhere 195
    11.4.3 Easy Tracking of Expenses 196
    11.4.4 Less Risk of Loss and Theft 196
11.5 Role of Emerging Technologies in Digital Payments 196
    11.5.1 Blockchain and Digital Payments 196
    11.5.2 Big Data Analytics and Digital Payments 198
    11.5.3 Social Media Analytics and Digital Payments 199
    11.5.4 Cloud Computing and Digital Payments 200
11.6 Economic Impact of Digital Payments in India 201
11.7 Discussion and Conclusion 201
11.8 Future Research Direction 202
References 202

11.1 INTRODUCTION

The world has changed the way it worked primarily due to the technological revolution. The consumers have witnessed the emergence of new options which not only save their time and money but also provide them much improved services, thanks to the emergence of advanced tools of information and communication (Vroman et al., 2015). The information and communication technologies along with the rise of the

Internet have opened new avenues of business, and digital mode of transactions is one of the most path-breaking innovations till date. Currency exchange has witnessed novelty of ideas in the digital sector over the past few years, but the contactless digital payment technology has changed the dynamics of the banking industry. The evolution of these payment methods is meant to transform the traditional digital wallet and the digitized user cards as they not only fail to provide the higher level of security due to lack of control on the hardware of the company but also necessitate the user to carry physical card which may make them redundant and replaceable with these new methods. Digital payments give the advantage to the user in the form of exchange of the money directly between the user's bank account and the merchant's bank account without any third-party involvement which is a win-win situation for both.

Payment gateways today provide the opportunity to consumers to utilize the apps and other modes of digital payments to make transactions on their platform. They help in making payments against electricity bills, cable operators, mobile bills, insurance companies, travel bookings, hotels, educational institutions, groceries shopping, etc. without the exchange of a single currency note. With developments in the field of technological innovations, the traditional system of making payments has made a shift toward digital platforms (Hromcová et al., 2014; Beijnen & Bolt, 2009; Bolt & Humphrey, 2007). Banks are constantly working toward increasing the efficiency of digital payment systems (Bouhdaoui & Bounie, 2012a). However, despite all these innovations and banks making substantial investment toward digital payments like debit cards, prepaid cards, contactless cards, and mobile payments, cash still remains the most preferred payment instrument (Arango-Arango et al., 2018; Bounie et al., 2016; Liu et al., 2015; Fung et al., 2015).

This chapter, thereby, attempts to identify the facilitators and inhibitors of digital payments in India and also explore the impact of emerging technologies on digital payments. The discussion and the investigation may provide the specific directions to the policy makers and the digital service providers that can further boost the adoption and usage of digital payments in India.

This chapter discusses the current status of digital payments in India in Section 11.2. The chapter then attempts to identify the factors that affect the usage of digital payments despite the associated numerous advantages. The inhibitors and facilitators of digital payments discussed in Sections 11.3 and 11.4 may eventually be major push or pull factors in drawing users of digital payments in the long run in India. In the following section, impacts of integration of digital payments with emerging technologies such as blockchain, big data analytics, social media analytics, and cloud computing are discussed. Finally, the economic impact of digital payments in India is also discussed.

11.2 DIGITAL PAYMENTS IN INDIA

The Digital India program is one of the most highly visioned programs of the government of India under Digital India movement. Various modes of digital payments are available to users in India such as UPI (Unified Payments Interface), mobile wallets, banking cards, mobile banking, AEPS (Aadhaar Enabled Payment System), point of sale, Internet banking, mobile banking, etc. The rise of digital payments in India got a boost post demonetization (Mohan & Kar, 2017). Moreover, the COVID-19 pandemic has only further enhanced the robust growth of digital payments in India, and Indian users are highly relying on UPI (Dalal, 2020). RBI and the government of India are already encouraging the use of digital payments in India (ETBFSI, 2020). The constant push from the government and rise in technological innovations have demarcated substantial growth of digital payments in India. The digital payment market in India is projected to reach INR 4,323.63 trillion by FY 2024 from INR 1,638.49 trillion in FY 2019 (KPMG Report, 2020). India is keen on developing the digital infrastructure, and the growth has been phenomenal over the past few years. Rural India is also being facilitated with 3G and 4G Internet network which will bring about digital revolution in India (RBI Report, 2020). Despite the encouraging trends of digital payments rise in India, cash transaction is still the preferred method. The government though is fully committed to the vision of Digital India; however, it can be envisioned only if the citizens of India embrace it in totality and see it as a potential replacement of cash transaction. Hence, there is a need to seek the Indian context-specific inhibitors and facilitators on the part of Indian citizens that may help bridge the gap in adoption of digital payments in India.

11.3 INHIBITORS OF DIGITAL PAYMENTS

There is no doubt in the fact that information technologies and financial market innovations have increased the prevalence of electronic payment systems; howsoever, they have not been able to replace the cash-based transactions yet (Immordino & Russo, 2017). Despite the continuous efforts of the government bodies and the banks together, there is still an inhibition in the minds of consumers in moving toward digital payments from the traditional methods of transactions. Many researchers have already attempted to get the nerve of consumers related to their reservations in terms of purchase which account as performance, financial, health, time, social acceptance, and also psychological constraints (Laroche et al., 2003; Lim, 2003). However, Cocosila and Trabelsi (2016) observed that users are more concerned about time, social, and psychological risks associated with contactless transactions as they may fear for their time being wasted due to the absence of direct contact with the server and nonacceptance from family while subscribing for a new service and may suffer from the anxiety while dealing with the advanced technology. It will be too early to talk of performance risks as digital payments are still in adoption stage. Financial risk as well as health risk may be ignored for obvious reasons as the online transactions do not involve any fee and certainly being contactless do not impact the health of the user. However, one important risk can be an additional type of risk which is expected to have a significant meaning for this research that is the privacy risk or data security risk. Users share their apprehension in dealing with online transactions due to the risks involved in getting their personal and confidential data leaked and becoming available in general (Yang et al., 2015; Grassie, 2007; Featherman & Pavlou, 2003). However, from the conversations with the users, they shared overspending risk as also the key inhibitor in the digital payments. This study, therefore, seeks to understand the role of all accumulated key inhibitors in the adoption of digital payments as follows.

11.3.1 SOCIAL RISK

Verkijika (2018) referred social influence as the key inhibitor of digital payments. The users do get influenced by the belief of users' near and dear ones about the usage of the new technological systems (Sharma & Sharma, 2019; Oliveira et al., 2014; Venkatesh et al., 2012). Social influence does play an important role in the adoption of digital payments as not only users consult their family and friends while deciding on usage of a new technology, but also, the positive opinions encourage them to adopt the same; however, a negative opinion may also discourage them from using the digital payments (Jaradat & Al Rababaa, 2013; Al-Louzi & Iss, 2011). Thereby, social risk may be considered as a strong inhibitor in the proliferation of digital payments in India.

11.3.2 PSYCHOLOGICAL RISK

Cocosila and Trabelsi (2016) observed that users have an overall feeling of anxiety while making an online payment. People at large are still not well versed about the technology of online payments and may get nervous about using it properly and correctly. As the digital payment modes are based on mobile phone, the digital payment system thereby may be somewhat difficult for nontechnical persons to learn. This sense of anxiety prohibits them from using the digital payment technology, and they continue doing the transactions in the traditional way. Therefore, psychological risk can be another key inhibitor in the proliferation of digital payments in a developing country like India.

11.3.3 TIME RISK

The users who are not early adopters of technology believe that online transactions may be more time-consuming in order to follow the process of making the payments, whereas they will not have to follow such payment method processes while making cash transaction. These users also fear that a wrongly done transaction may further involve the extra time in canceling the same and proceeding for the new. Thereby, time risk can be considered as an important inhibitor of the proliferation of digital payments in India.

11.3.4 DATA SECURITY RISK

Digital payment network presents its own set of security risk which is not the case in a cash-based transaction as the identity of the payer is not required, whereas electronic payments present risks of fraud or error due to unauthorized transactions which may result in data theft as well (Morse, 2018). It is, therefore, obvious that data security risk becomes an important inhibitor in the adoption of contactless payments (Bagadia & Bansal, 2016; Yu et al., 2012). The data hackers can hack the servers of the bank or the e-Wallet and may easily get access to the personal information of the account holder for stealing money from the account. The other third parties may also use the users' personal data for their various other tasks. Thereby, data security risk is also another dimension inhibiting proliferation of digital payments in India.

11.3.5 Overspending Risk

Users carry limited cash in their physical wallet and therefore cannot spend beyond what they have in their wallet at the moment, whereas in the case of digital payment modes, the users have all their money with them always, and that may result in overspending as the human beings by nature get tempted toward the new offerings in the market. So, the users feel that they will not carry any overspending risk if they do not subscribe the facility of online transactions at the first place. Thus, overspending risk should also be explored as the key inhibitor in digital payments in India.

11.4 FACILITATORS OF DIGITAL PAYMENTS

Despite the inhibitions toward technology usage, consumers prefer the convenience of transacting through online payments and have started realizing the benefits of digital payments. Consumers experience the expansion in liquidity due to the electronic access to funds, and this also builds a relationship of trust within payment networks (Morse, 2018). The facilitators thus can be understood as the perceptions of consumers toward available resources that ease the use of online transactions (Chopdar et al., 2018). The key facilitators of digital payments therefore can be listed as follows.

11.4.1 Easy and Convenient to Use

Way back, Yiu et al. (2007) observed that users are more likely to adopt Internet banking if they find it easy to use. Digital payments are certainly easy and convenient to use. In fact, the ease of doing financial transactions is one of the biggest motivators to go digital. Users do not need to take along loads of cash anymore. They can make any amount of payment without carrying a physical wallet. They can also experience the convenience of not carrying various currency denominations. With digital payment facility, banking services are not only available at the doorstep but also available 24 × 7 providing the users the most convenient way of doing any financial transaction irrespective of time and place. Thereby, the ease and convenience associated with the digital banking should be a key facilitator for the usage of the same in India.

11.4.2 Enabled Transaction from Anywhere

With digital payment modes, one can pay from anywhere anytime. The payer need not be present at the place and also can make the payment on behalf of the near and dear ones in the time of need. Electronic payments by allowing the transmission of value by electronic means save time as well as cost associated with traditional payment systems especially in the case of long distances (Morse, 2018). Thus, e-transactions provide the opportunity to save time and money (Kim et al., 2009). This facility is particularly useful in emergencies such as hospital where a service might be affected due to delay in payment. With digital payments, the users do not have to follow any timeline and place restriction or ensure self-availability for any service. Thus, this comfort of making transactions irrespective of time and distance should be an important facilitator of digital banking usage in India.

11.4.3 Easy Tracking of Expenses

The users no longer need to maintain a written record of their spending if they are making digital payments. The transactions are automatically registered in the e-passbook or e-wallet, and the users may track their dealings at any point of time. The readymade list of all the expenses may prove to be handy to the user for the comparison of their budget with the actual expenditure. There are many applications such as AndroMoney, Dollarbird, Goodbudget, etc. associated with the digital banking that automatically matches the recurring expenses and goals and help the users to make smart decisions (Sharf, 2016). Consequently, the facility of tracking of expenses may be a strong facilitator of digital banking usage in India.

11.4.4 Less Risk of Loss and Theft

Digital payments are less risky if used wisely. The cash stolen can be used by anyone, whereas no one can use ones' digital money without MPIN, PIN, or the fingerprint of the actual user in case of theft of a mobile phone or a laptop/tablet, and the user can simultaneously suspend the wallet account to prevent misuse in case of theft or loss. Moreover, banks offer strong information security policies, procedures along with advanced technology in the name of unique user IDs and passwords, security questions, security key, etc. to guard against the misuse of digital transactions by the unknown person. Thereby, the very nature of cashless transactions of digital banking combined with the security features offers less risk to the users and may become a facilitator of digital banking usage in India.

These identified inhibitors as well as facilitators divulge that there is a definite need to educate the existing as well as the prospective users about the facilitators of digital payments and curb the inhibitors in order to enhance the digital participation. Apart from educating and motivating the users, the policy makers and the service providers also need to exploit the emerging technologies in order to make a smooth transition from the cash transactions toward the cashless ones.

11.5 ROLE OF EMERGING TECHNOLOGIES IN DIGITAL PAYMENTS

The rising technologies not only reduce the cost of digital infrastructure but also contribute to the growth of digital payments. Banks and financial institutions are adopting these technologies for the faster transactions and better usage of data. The integration of few such technologies with digital payments and the subsequent impact are described below.

11.5.1 Blockchain and Digital Payments

Blockchain is a new technology which has potential to change the world in the next 20 years as Internet has changed the world in the past 20 years. Initially, blockchain was associated with crypto currencies and banking sector, and a white paper with the title "Bitcoin: A Peer-to-Peer Electronic Cash System" was published by Satoshi Nakamoto in 2008. Bitcoin was formally introduced to the end users in the year 2009. Satoshi Nakamoto is considered as the father of blockchain technology in the digital world. Blockchain is defined as a decentralized database or a ledger, in other words, a chain of multiple blocks that thwarts stakeholders from producing wrong information or records. A formal definition of blockchain is proposed by Beck et al. (2017) as "a distributed ledger technology in the form of a distributed transactional database, secured by cryptography, and governed by a consensus mechanism". In simple words, blockchain is being recognized as a chain of digital records. Buterin (2014) emphasized that the meaning of record is not just a record here, it is a software program implemented in blockchain without any risk and fraud. A simple example of blockchain is Wikipedia. The important attribute of blockchain is to remove trust component from individuals and put trust components in the mathematical modeling which ensures minimum possible error. Tapscott and Tapscott (2017) investigated the impact of blockchain technology on finance and summarized five key principles of blockchain technology and we are adopting these principles in this chapter and presenting in Figure 11.1.

The integration of digital payments with blockchain technology will increase trust of individuals and organizations in online payments. Luther (2016) argued that the "share of electronic transactions will continue to increase". In the past couple of years, it has been witnessed in India that the intention to shop online is on rise as digital payments are considered more convenient in comparison to traditional shopping methods. As per Statista (2020), the mobile penetration rate is expected to reach 61% by the year 2025 in Asia, and India would be a major contributor of that.

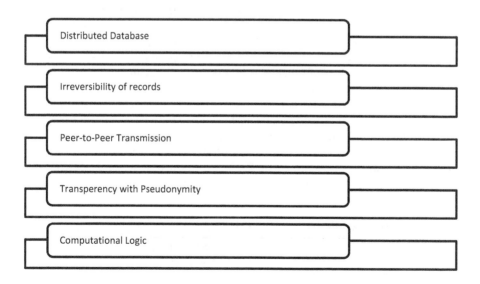

FIGURE 11.1 Five principles of blockchain technology. (Adapted from Tapscott & Tapscott, 2017.)

Luther (2016) further argued that blockchain technology will be accepted soon in the digital payments. In the present scenario, the transaction processing cost using blockchain technology is relatively less than traditional methods. Some big organizations have taken initiatives to employ blockchain technology. For example, NASDAQ has already made it public that blockchain-based digital ledger technology will be used to manage equities.

11.5.2 BIG DATA ANALYTICS AND DIGITAL PAYMENTS

Big data is considered as one of the latest technological innovations or smart revolution in the past one decade which has forced many big companies to change their strategy from traditional decision-making to data-driven decision-making. In the extant literature, researchers claim without any doubt that the big data is going to reshape the economy of any country. When we think of big data, the first thing that comes to our mind is size. However, other characteristics of big data are equally important. The big organizations such as Google, Amazon, Facebook, and Walmart have been using big data and gaining competitive advantage in the complex environment. Many organizations have proposed their own definitions of big data. Gartner, Inc. defines big data as follows: "Big data is high volume, high velocity, and high variety information asset that demands cost-effective, innovative forms of information processing for enhanced insight and decision-making. The 3V's of big data were firstly coined in 2001 by Gartner Inc. analyst Doug Laney (Gartner n.d.). Later, three big companies, namely, IBM, SAS, and Oracle, proposed three more dimensions of the big data such as *veracity*, *variability*, and *value*. The detailed discussion on these six key dimensions of the big data can be followed in the article written by Gandomi and Haider (2015).

In the digital world, digital payments are being promoted and motivated for secure and faster transactions. As a result, a large amount of data is being created and stored in the larger repositories of financial institutions across the globe. Such type of data possesses all characteristics of big data. An appropriate big data analytics can be employed to extract useful information. The process of extracting useful information from big data can be divided in five stages as suggested by Labrinidis and Jagadish (2012) and given in Figure 11.2. The five stages of big data analytics are summarized in two sub-stages, namely, data management and analytics.

The prominent types of big data analytics such as text analytics, audio analytics, and predictive analytics can be employed on the digital payments data set to gain competitive advantage. The text analytics is also termed as text mining which is primarily a useful approach to extract information from the data available in the text

Data Management	Analytics
• Storing data • Extraction and cleaning data • Integration, aggregation, and representation	• Appropriate model and analysis • Interpretaion

FIGURE 11.2 Stages for big data analytics. (Adapted from Labrinidis & Jagadish, 2012.)

form. The huge amount of data is being generated through news articles, blogs, and logs of call centers from banking industry, and text analytics provides summaries of large text data to support evidence-based decision-making in banking as well as other industries. Audio analytics is another important type of big data analytics to extract useful information from the large set of audio data which is unstructured in nature. Many call centers in the banking sector receive calls from customers related to digital payments queries and are stored in repositories. Audio analytics helps in extracting useful information from such repositories to develop effective and successful strategies to attract more users to use digital payment services in the banking sector. Predictive analytics is a group of analytical tools such as regression techniques and machine learning techniques, which attempts to predict future events based on past and present data sets. The regression techniques include logit regression, probit regression, etc., whereas machine learning techniques include neural networks, random forest, etc. The predictive analytics can be used to predict the behavior of customers toward digital payment usage on the basis of the past data set.

In the Asia-Pacific region, decision makers are spending a huge amount of money on the digital infrastructure, and as a result, a huge amount of data is being generated in almost all business domains and banking sector in particular. This is the right opportunity to employ big data technologies and analytics in the banking sector for advanced and data-driven decision-making. The extraction of useful information from the available data sets is one of the strategic priorities for banking institutions. Big data technologies and analytical capabilities help financial institutions to understand online as well as offline behavior of digital payment users and provide insights to deliver customized and targeted service quality to the end users. As it is evident from multiple applications of the big data technologies, these technologies are very powerful in nature and have capabilities to handle some serious cyber financial crimes such as money laundering.

11.5.3 Social Media Analytics and Digital Payments

The advances in the social media technology have paved the way for the access of a huge amount of data through online social media platforms. The data generated through social media is providing competitive advantages in the form of analytics to leading industries particularly in banking, real estate, healthcare, etc. Social media analytics is defined by Gandomi and Haider (2015) as "the analysis of structured data and unstructured data from social media channels". In the present context, social media provides a number of online platforms such as Facebook, Twitter, WordPress, Reddit, Delicious, Instagram, Yelp, and many more. The structured data means that data is presented in tabular form which is available in Excel sheets or relational databases. Cuckier (2010) reported that the available structured data in all forms of the literature are less than 5%. The data available in the multiple formats such as text, audio, video, and images are known as unstructured data. The unstructured data do not follow structural organization of data. Social media is not considered as a new term in the extant literature, whereas social media analytics is in its early stage in the literature. The main attraction of the social media analytics lies in the data-centric methodology adoption to extract useful information from the user-generated structured and unstructured data sets. Gandomi and Haider (2015) further stressed that the content generated by users namely opinions, reviews, feedbacks, audios, etc. and association among key stakeholders such as users and industries are the main contributors of the generation of information. McKinsey Global iConsumer Research (2012) highlights the importance of digital connection among younger generation under continuously changing global environment. Younger generation performs the following tasks more in relative comparison to elderly people:

1. Possess Internet-enabled high-end mobile phones
2. Spend significant time on social media, video calls, and others
3. Frequently buy items on the Internet

There is a problem of poor utilization of social media platform with no expert support. The emergence of new tools and techniques should be complemented by social media-generated data. Social media analytics provides valuable information to decision makers if monitored and employed effectively and strategically. For example, if customers are not happy with digital payment services, they will generally go to social media and start sharing their views which spread a negative image of the services. This type of negative discussions can be checked well in advance with the support of social media analytics and provide valuable feedback about services to the decision makers.

11.5.4 Cloud Computing and Digital Payments

The growth of cloud computing in the past decade is attributed to the way it changed information technology services. The prices of IT infrastructure are decreasing, whereas the cost to maintain complex IT infrastructure is increasing. The cloud computing services provide a feasible and effective solution to organizations with minimal resources and aspire for greater results in the competitive environment. A recent survey conducted in six data centers at big organizations revealed interesting facts that utilization of their server is up to 10–30% of their computing power, and utilization of desktop computers is less than 5% of the available capabilities (Marston et al. 2011). Bhattacherjee and Park (2014) defined cloud computing as "a new paradigm of computing where users employ third party, Internet-hosted software, applications, data storage, and computing services for their computing needs rather than applications on their local computers". In the present context, it is almost essential for all organizations to use cloud computing services. The cloud computing services reduce cost of new organizations with relatively less resources, which intend to take competitive advantage of emerging tools and technologies normally available with big organizations. The cloud service providers generally deliver in three commonly accepted models namely software as a service (SaaS), platform as a service (PaaS), and infrastructure as a service (IaaS). These three delivery modes are also known as three different layers in the cloud computing architecture. The commonly used delivery model is SaaS, and popular examples are Gmail and Facebook, and the examples of PaaS are Google App Engine and Rackspace Cloud Site. The IaaS comprises Terremark Worldwide and Amazon storage services. For detailed discussion about cloud computing, benefits, and applications, readers may refer to Marston et al. (2011).

The banking industries in some leading countries are taking advantage of cloud computing services for creating a flexible environment as per the requirements of the customers (Asadi et al., 2017). Banks with limited resources and even newly launched banks opt for cloud services to save money by paying just for the services needed. In case of mobile commerce and Internet commerce, users are using digital payment options. The data generated using digital payments are being stored in the cloud repositories. The big data analytics discussed in the previous section are being employed to extract customer behavior and other useful trends using real time data set provided by cloud services to targeted marking purposes to enhance digital payments across the world. Further, financial institutions use services provided by cloud computing for customer segmentation.

11.6 ECONOMIC IMPACT OF DIGITAL PAYMENTS IN INDIA

The advent web 2.0 and advanced mobile technologies are being leveraged by users and organizations for effective utilization of available resources. Digital payments facilitate individual users and organizations to save time efforts and financial resources and increases flow of money in the markets to unleash tremendous potential of entrepreneurial activities in the country. The greater acceptance of digital payments will assist financial institutions in minimizing their expenditure by reducing spending on brick and mortar branches and hiring a lower number of new employees as customers' visits to banks will reduce significantly. Furthermore, digital payments provide greater opportunities to banks in expending their customer base as in the case of India, Internet and other required infrastructure are available in every part of the country. In the case of India, the government is spending a substantial amount of the budget on the development of digital infrastructure. As a result, the usage of digital payments will grow substantially among individual users and also among small as well as big organizations. The Nandan Nilekani Committee (2019), constituted by RBI, reported that there is a huge potential for the volume of growth in digital payments in India and gave a vision to increase the digital transaction value to 1,500% of GDP by FY 2021 from 769% in 2018.

11.7 DISCUSSION AND CONCLUSION

The objective of this chapter was to review, explore, and reflect on the various dimensions of digital payment in India. The digital payments will have a long-lasting impact on the economy of India, resulting in various advantages to all stakeholders in the country such as individuals, public and private organizations, and business, among others. This chapter discourses various facets of challenges as well as opportunities created by the proliferation of digital payments from the perspectives of a developing country. This chapter further explored the impact of emerging technologies such as blockchain, big data analytics, social media analytics, and cloud computing on the effective and strategic use of digital payments. In addition, the economic impact of digital payments in India has also been discussed. Finally, this chapter provides useful information to decision makers in the banking industry and researchers for developing effective and successful strategies to receive substantial return on the investment in developing digital infrastructure by the government in India.

11.8 FUTURE RESEARCH DIRECTION

The current study is an exploratory study which attempts to identify the facilitators as well as inhibitors of digital payments in India. However, this study provides a number of impactful and actionable insights in the area of digital payments. This exploration may help the researchers in devising the suitable theoretical model of digital payment adoption in India. The future researchers can extend the study with the empirical investigation of the digital payment inhibitors and facilitators. The researchers may also substantiate the impact of the emerging technologies with the policy makers and the digital service providers. Such further studies may provide a deeper understanding of digital payment adoption in India.

REFERENCES

Al-Louzi, B. & Iss, B. 2011. Factors influencing customer acceptance of M-commerce services in Jordan. *Journal of Communication and Computer*, 9, pp. 1424–1436.

Arango-Arango, C. A., Bouhdaoui, Y., Bounie, D., Eschelbach, E. & Hernandez, L. 2018. Cash remains top-of-wallet! International evidence from payment diaries. *Economic Modelling*, 69, pp. 38–48.

Asadi, S., Nilashi, M., Husin, A. R. C. & Yadegaridehkordi, E. 2017. Customers perspectives on adoption of cloud computing in banking sector. *Information Technology and Management*, 18(4), pp. 305–330.

Bagadia, P. & Bansal, A. 2016. Risk perception and adoption of mobile banking services: A review. *IUP Journal of Information Technology*, 12(1), pp. 52–71.

Beck, R., Avital, M., Rossi, M. & Thatcher, J. B. 2017. Blockchain technology. *Business & Information Systems Engineering*, 59, pp. 381–384.

Beijnen, C. & Bolt, W. 2009. Size matters: economies of scale in European payments processing. *Journal of Banking & Finance*, 33(2), pp. 203–210.

Bhattacherjee, A., & Park, S. C. 2014. Why end-users move to the cloud: A migration-theoretic analysis. *European Journal of Information Systems*, 23(3), pp. 357–372.

Bolt, W. & Humphrey, D. 2007. Payment network scale economies, SEPA, and cash replacement. *Review of Network Economics*, 6(4), pp. 453–473.

Bouhdaoui, Y. & Bounie, D. 2012a. Efficient payments: how much do they cost for the central bank? *Economic Modelling*, 29(5), pp. 1579–1584.

Bounie, D., François, A. & Waelbroeck, P. 2016. Debit card and demand for cash. *Journal of Banking & Finance*, 73(C), pp. 55–66.

Buterin, V. 2014. Ethereum white paper. Retrieved October 23, 2020, from http://www.theblockchain.com/docs/Ethereum_white_papera_next_generation_smart_contract_and_decentralized_application_platform-vitalik-buterin.pdf.

Chopdar, P. K., Korfiatis, N., Sivakuma, V. J. & Lytras, M. D. 2018. Mobile shopping apps adoption and perceived risks: A cross-country perspective utilizing the unified theory of acceptance and use of technology. *Computers in Human Behavior*, 86, pp. 109–128

Cocosila, M. & Trabelsi, H. 2016. An integrated value-risk investigation of contactless mobile payments adoption. *Electronic Commerce Research and Applications*, 20, pp. 159–170.

Cukier, K. 2010. Data, data everywhere: A special report on managing information. *The Economist*, 394, pp. 3–5.
Dalal, V. 2020. Where India leads in digital payments, and where it lags, pp. 428–438. Retrieved October 18, 2020, from https://www.livemint.com/news/india/where-india-leads-in-digital-payments-and-where-it-lags-11603181605154.html.
ETBFSI. 2020. RBI pushes digital payments in the time of COVID-19. Retrieved November 11, 2020, from digital payments: RBI pushes digital payments in the time of COVID-19, BFSI News, ET BFSI (indiatimes.com)
Featherman, M. S. & Pavlou, P. A. 2003. Predicting e-services adoption: A perceived risk facets perspective. *International Journal of Human-Computer Studies*, 59(4), pp. 451–474.
Fung, B., Huynh, K. P. & Stuber, G. 2015. The use of cash in Canada. *Bank of Canada Review*, 2015, pp. 45–56.
Gandomi, A., & Haider, M. 2015. Beyond the hype: Big data concepts, methods, and analytics. *International Journal of Information Management*, 35(2), pp. 137–144.
Gartner (n.d.) Information technology gartner glossary. Retrieved October 14, 2020 from https://www.gartner.com/en/information-technology/glossary/big-data
Grassie, K. 2007. Easy handling and security make NFC a success. *Card Technology Today*, 19(10), pp. 12–13.
Hromcová, J., Noz, F. J. C.-M. & Utrero-González, N. 2014. Effects of direct pricing of retail payment methods in Norway. *Economic Modelling*, 37. doi:10.1016/j.econmod.2013.11.036.
Immordino, G. & Russo, F. R. 2017. Cashless payments and tax evasion. *European Journal of Political Economy*, 55(C), pp. 36–43.
Jaradat, M.-I. R. & Al Rababaa, M. S. 2013. Assessing key factor that influence on the acceptance of mobile commerce based on modified UTAUT. *International Journal of Business and Management*, 8(23), pp. 7–11.
Kim, D. J., Ferrin, D. L. & Rao, H. R. 2009. Trust and satisfaction, two steppingstone for successful ecommerce relationships: A longitudinal exploration. *Information Systems Research*, 20(2), pp. 237–257.
KPMG Report. 2020. Impact of COVID-19 on digital payments in India. Retrieved November 11, 2020, from Impact of COVID-19 on digital payments in India (assets.kpmg)
Labrinidis, A. & Jagadish, H. V. 2012. Challenges and opportunities with Big data. *Proceedings of the VLDB Endowment*, 5(12), pp. 2032–2033.
Laroche, M., Bergeron, J. & Goutaland, C. 2003. How intangibility affects perceived risk: The moderating role of knowledge and involvement. *Journal of Services Marketing*, 17(2), pp. 122–140.
Lim, N. 2003. Consumers' perceived risk: Sources versus consequences. *Electronic Commerce Research and Applications*, 2: pp. 216–38
Liu, J., Kauffman, R. J. & Ma, D. 2015. Competition, cooperation, and regulation: Understanding the evolution of the mobile payments technology ecosystem. *Electronic Commerce Research and Applications*, 14(5), pp. 372–391.
Luther, W. J. 2016. Bitcoin and the future of digital payments. *The Independent Review*, 20(3), pp. 397–404.
Marston, S., Li, Z., Bandyopadhyay, S., Zhang, J. & Ghalsasi, A. 2011. Cloud computing— The business perspective. *Decision Support Systems*, 51(1), pp. 176–189.
McKinsey Global iConsumer Research. 2012. The young and the digital: A glimpse into future market evolution. Retrieved October 10, 2020, from https://www.mckinsey.com.
Mohan, R. & Kar, A. K. 2017. Demonetization and its impact on the Indian economy-insights from social media analytics. In *Conference on e-Business, e-Services and e-Society* (pp. 363–374). Springer, Cham.
Morse, E. A. 2018. From Rai stones to blockchains: The transformation of payments. *Computer Law & Security Review*, 34, pp. 946–953.

Nandan Nilekani Committee. 2019. Deepening of digital payments. Retrieved October 28, 2020, from Committee Reports (prsindia.org).

Oliveira, T., Faria, M., Thomas, M. A. & Ales, P. 2014. Extending the understanding of mobile banking adoption: When UTAUT meets TTF and ITM. *International Journal of Information Management*, 34, pp. 689–703.

RBI Report. February 24, 2020. Assessment of the progress of digitisation from cash to electronic. Retrieved November 15, 2020, from https://m.rbi.org.in/Scripts/PublicationsView.aspx?id=19417.

Sharf, S. 2016. 12 Free Apps to track your spending and how to pick the best one for you. Retrieved November 10, 2020, from https://www.forbes.com/amp/s/sites/samanthasharf/2016.

Sharma, S. K. & Sharma, M. 2019. Examining the role of trust and quality dimensions in the actual usage of mobile banking services: An empirical investigation, *International Journal of Information Management*, 44, pp. 65–75.

Statista. 2020. Mobile internet user penetration in APAC 2018–2025. Retrieved November 19, 2020, from https://www.statista.com/statistics/201232/forecast-of-mobile-internet-penetration-in-asia pacific/#:~:text=In%20the%20Asia%20Pacific%20region, in%20 the%20Asia%20Pacific%20region.

Tapscott, A., & Tapscott, D. 2017. How blockchain is changing finance. *Harvard Business Review*, 1, 2–5.

Venkatesh, V., Thong, J. Y. & Xu, X. 2012. Consumer acceptance and use of information technology: Extending the unified theory of acceptance and use of technology. *MIS Quarterly*, 36(1), pp. 157–178.

Verkijika, S. F. 2018. Factors influencing the adoption of mobile commerce applications in Cameroon. *Telematics and Informatics*, 35, pp. 1665–1674.

Vroman, K., Arthanat, S. & Lysack, C. 2015. "Who over 65 is online?" Older adults' dispositions toward information communication technology. *Computers in Human Behavior*, 43, pp. 156–166.

Yang, Q., Pang, C., Yen, D. C. & Tarn, J. M. 2015. Exploring consumer perceived risk and trust for online payments: An empirical study in China's younger generation. *Computers in Human Behavior*, 50, pp. 9–24.

Yiu, C.S., Grant, K. & Edgar, D. 2007. Factors affecting the adoption of internet banking in Hong Kong—Implications for the banking sector. *International Journal of Information Management*, 27, pp. 336–351

Yu, A. W., Prybutok, V. R., Koh, C. E. & Hanus, B. 2012. A nomological model of RFID Privacy concern. *Business Process Management Journal*, 18(3), pp. 420–444.

12 Cryptocurrency
Perspectives, Applications, and Issues

Abhishek Sharma, Ayush Srivastava, and Deepika Dhingra
Bennett University

CONTENTS

12.1 Introduction 205
12.2 Background 206
12.3 Research Objective 207
12.4 Perspectives 207
    12.4.1 Neoclassical Finance and Economics 207
    12.4.2 Behavioral Economics..208
    12.4.3 Socioeconomic Perspectives 210
12.5 Applications of Cryptocurrencies 211
    12.5.1 Travel 211
    12.5.2 Education 211
    12.5.3 Banking and Financial Services...212
    12.5.4 Wealth Management 212
    12.5.5 Wide Usage and Easy Accessibility 212
12.6 Legal Issues Associated with Cryptocurrency 212
12.7 Environmental Issues Associated with Cryptocurrency 214
12.8 Scams 215
12.9 Implications of Cryptocurrency 215
12.10 Scope of Growth Opportunities for Cryptocurrency 216
12.11 Conclusion 218
References 218

12.1 INTRODUCTION

It goes without saying that the era of communication technologies and data has led to many golden opportunities in almost every sector. One area which is hugely benefited is the financial sector. An increase in the number of internet users has led to the activation of the virtual world notion and created an altogether fresh business phenomenon. Hence, diversification of transactions, trading, and currencies is continuously rising. The most innovative form in the financial system in the 20th century is cryptocurrency. A cryptocurrency is known as any medium of exchange that can be used to purchase goods and services and transfer money without any formalities.

The applications of cryptocurrency are profound. It is present in every sector be it education, travel, banking, and financial services. For instance, cryptocurrencies have changed the way people conduct bank transactions. It not only provides services that all the traditional banks do but also provides transparency, a more efficient way of managing transactions and security to its users. Cryptocurrency results in either free or low cost in making transactions.

The unique factor about cryptocurrency is deregulation, which implies the absence of central authority that can monitor or regulate the supply of cryptocurrency tokens. All transactions are recorded on software and then added to public ledger, making it more accessible for people. Many regulatory bodies and experts think that all the advantages which this type of currency offers come with a cost and are worried about its usage in unlawful activities like money laundering, tax evasion, and terrorist financing. The problem is noteworthy, and even though the full scale of misuse of digital currencies is unknown, its market value is expected to exceed in the coming future by almost € 7 billion worldwide. For specific peculiarities of cryptocurrencies, a ban could be considered. The feature of intractability comes to mind, which according to many countries is the root cause of all illegal and illegitimate activities, and now, a lot of questions come to everybody's mind. Is such a degree of anonymity truly wanted? Can it be evaded? Would a ban eliminate criminals' activities? In any situation, imposing a ban should only be focused on specific aspects that can curb the illicit use of cryptocurrency but not on the whole system.

Cryptocurrencies have been alluring entrepreneurs, regulators, investors, and the public in general. Lately, there have been discussions about cryptocurrencies triggering substantial changes in prices of various commodities, and many stakeholders claim that the market for cryptocurrencies is a bubble without any foundational value for which the concerns about evasion of regulatory and legal omission should not be ignored. These matters have led to warning calls for increased regulation.

12.2 BACKGROUND

In ancient times, people did not use to have any medium of exchange, like currencies or any derivatives to fulfil their necessities. The trade of goods and services happened through the barter system. There was a major flaw in the barter system. Suppose there are two people, Mr. A and B, Mr. B wants to buy wheat in exchange for rice, but Mr. A is offering wheat but only in exchange for apples with wheat, so in this scenario, no party's need is fulfilled. As time had passed, gold had become a major mode of exchange and that turned out to be a little better than barter exchange, but it also had some drawbacks too. In 1600 AD, the currency notes and coins came to light. Back then, currency notes and coins were widely accepted because they were the only means of payment which was very convenient to use and became very popular in no time. This is how modern currency came into existence. Modern currency not only includes paper notes or coins but also includes digital valets, credit cards, and debit cards. Generally, all the modern currency is controlled by the central bank of that country. But there exist few drawbacks which haven't been resolved. For example, suppose that a person X is doing an online transaction and the bank does not approve of the transaction due to some technical problems like system failure at the bank or their accounts have been hacked. So, this is why the future of currency relies on a digital currency like cryptocurrency.

Cryptocurrency in the mid-1990s was just a theoretical concept in which many mathematics and computer science principles have been applied; the objective was to eliminate the drawbacks faced by the users of fiat currencies. Back in the 1980s, a cryptographer called David Chaum invented an algorithm named "blinding" which was central to modern web-based encryption. The algorithm focused on unalterable information, secured information, and most importantly laid all the groundwork which was needed to carry the future of digital currency in the future. Later, Chaum's concept got popular and Schaum's founded a company called DigiCash, a Netherlands for profit company that produced units of currency based on blinding algorithms, but Chaum's company had a monopoly on controlling the money supply of this currency. DigiCash was directly dealing with individuals, so after seeing the work done by Chaum's company and to protect the interest of the general public, the central bank of Netherland quashed this idea. Later, an associate of Chaum released a cryptocurrency called Bitgold which used blockchain. Bitgold never gained popularity because of very few transactions happening on the software.

12.3 RESEARCH OBJECTIVE

This paper discusses the evolving phenomenon of cryptocurrency. The main objective of this paper is to extend the prevailing awareness encompassing cryptocurrencies, which exemplifies revolutions and high-tech change and may seem to be a rewarding form of fundraising for small firms, focusing on the environment, sustainability, and social responsibility. To attain this main objective, the followings are other subobjectives of the study. The growth of cryptocurrencies' value in the market and the increasing acceptance around the globe open several trials and worries for business and industrial economics. The paper also attempts to study two significant issues such as legal and environmental faced by cryptocurrency.

Cryptocurrencies such as Bitcoin, Ether, and others are a hot commodity in online trading, and it is probable for a keen investor to reap large profits. But the potential of speedy riches can vizor people to the risks and permit staves to bait them into scams. The second last section of the paper sheds light on the various scams related to cryptocurrency. The last leg of the paper mentions the potential of cryptocurrency and amp, the implications of the same.

12.4 PERSPECTIVES

12.4.1 Neoclassical Finance and Economics

Cryptocurrency being adopted as a payment method and as a financial asset has changed the way people conduct financial transactions globally. It was proved with evidence that the main logic behind buying a cryptocurrency is the predictable nature of assets (Glaser et al. 2014). Hybrid financial securities like CFDs (derivative products) and ETNs (exchange-traded notes) are available, imitating the performance of Bitcoin for a larger set of investors to expand the speculative investment opportunities. Evaluation of cryptocurrency as a financial asset seems to be a sensible idea.

Several studies conducted a cross-sectional analysis of cryptocurrencies' return. Urquhart (2016) displayed that earnings related to Bitcoin do not trail a random walk; they are derived based on the conclusion that the market of Bitcoin displays the gravity of ineffectiveness, predominantly in its initial phase. Corbet et al. (2018) examined the association between several financial assets and the return of three different cryptocurrencies, signifying nonexistence of relation between cryptocurrencies and other financial assets. Tsyvinski and Liu (2018) examined if crypto pricing shows contrast to other stocks or not. In most of the cryptocurrency, events in economic factors, which holds conventional implication for other assets contribute a negligible or no role. The paper denies cryptocurrency as any substitute for money or holding value like gold but emphasizes that they have their asset class.

Corbet et al. (2019) summarized cryptocurrency as a beneficial and legitimate payment system and the credible option of the investment asset class. The distinguishing feature of cryptocurrency from other classic financial assets gives cryptocurrency another dimension that offers benefits of diversification to investors with a short-range asset limit. Both Bouri et al. (2017) and Baur et al. (2018) discovered that cryptocurrency is appropriate for divergence as its yields don't correlate, with most of the other primary asset classes. Intriguingly, providing accurate and useful proved to be the prominent use of Bitcoins as speculative assets.

Fantazzini and Zimin (2020) discussed the risk of holding cryptocurrencies. The prices of cryptocurrency fluctuate owing to revelation scam or supposed cyber hacking or other various factors. Because of the crash in Coinbase digital exchange on June 26, 2019, Bitcoin price fell for greater than 10% of its price in a short time frame. In case of any such events and its consequence, a crypto coin may become unsalable, and its value may decline drastically. There was a set of models which were tested on 42 digital coins proposed by Fantazzini and Zimin.

12.4.2 Behavioral Economics

This section elucidates market sensation that functions in contradiction to neoclassical finance forecasts, from the outlook of immeasurable risk or uncertainty. In these circumstances of cryptocurrencies, there may be two reasons in this kind of improbability: (1) the technology may be complicated and not feasible for naïve brokers and (2) imprecise and unclear value of cryptocurrency. Dow and da Costa Werlang (1992) demonstrated that under gloomy confusion aversion, unpredictability about foundational leads to no trading in financial markets, hitherto this doesn't appear to be appropriate for cryptocurrencies. Though returns of cryptocurrency demonstrate high instability, trade volumes are noteworthy and important. In Caballero and Krishnamurthy (2008), cause such as "flights to quality" in markets, if appropriately applied to cryptocurrencies, might also elucidate about the crashes we lately witnessed.

Gaining and analyzing information is critical and can help in taking decisions to contain or lower unpredictability. FigàTalamanca (2020) focused on how the potency of the internet searches for cryptocurrency-related keywords considerably impacts cryptocurrency volatility (but not return); this influence disappears when one takes control of "pertinent happenings". These "pertinent happenings" are declarations of limitations/prohibitions on usage or broadening of the cryptocurrency market. Whether we may remain unsure of what users find on the internet when cryptocurrency-related terms are searched, the actions give us a hint of the kind of information and evidence that is essential for investment decisions in cryptocurrency and henceforth for a price assessment.

There may be numerous other reasons, as to why neoclassical predictions fail, Shiller (2003) observes that people or investors who participate in the market are humans, and they can make unreasonable methodical errors that may be different from the assumption of prudence. These errors impact values and earnings of assets, thereby generating inefficiencies in the market. Studies in behavioral economics stress upon unproductivity like no or extreme reactions to information, because of trends and in an extreme case of bubbles and crashes. These responses to market information have been credited to restricted attention on an investor, robust decision-making, imitation of other investors and noise trading, description of several references in Kahneman and Tversky's (1979) prospect theory, which suggested that people who make decisions evaluate consequences from the perspective of their present endowment and "review" possibility of results when deciding (chiefly giving emphasis on probabilities of negative results and not emphasizing on positive ones). The loss aversion led Shefrin and Statman (1985) to frame the 'disposition effect' in investment decisions; investors have a mindset of selling those assets that gain value too early and keeping those assets that lose value for too long.

Some key points which differentiate the digital-currency market and those which are noninstitutional investors are fluctuating returns and uncertainty in the principal quotient. Keeping in view these situations, behavioral preferences must be more prominent than in standard asset markets. Haryanto et al. (2020) studied the herding behavior and distribution effect in the cryptocurrency empire by scrutinizing the trading behavior at a crypto exchange. It concluded an opposite temperament effect in uptrend periods where the value of all cryptocurrencies surged while an optimistic disposition effect was noticed in bearish times. It was also found that in various market conditions, the herd follows the market.

The opposite temperament effect in the uptrend price market implies users reveal increased confidence and predicted returns to surge more, which is constant with the exponential price growth in a bubble in the nonappearance of clear foundational value. This lack of predictability concerning the foundational value is backed by irregular herding behavior as well. When the value increases in an optimistic market, participants look at other investors in the market and observe if they have the same opinion or not. Similarly, investors also observe others in participants in the pessimistic market (2020) to estimate the risk of default in cryptocurrency. They asserted that the analysis of conventional risk to cryptocurrencies and differentiating between market and credit risk is related to changes in prices of other assets.

12.4.3 Socioeconomic Perspectives

Experts often educate people about cryptocurrencies, which are not free from cheating and scams. For instances, billions in Bitcoin from the interface in Japan Mt. Gox in 2014 and $50 million in Ether during an event called Decentralized Autonomous Organization (DAO) brawl in 2016 were robbed. Böhme et al. (2015) summarized findings and data displaying that at the initial phase of the Bitcoin period, many dealings were used for the purchase of drugs. Foley et al. (2019) claim that approximately 46% of Bitcoin dealings are related to unlawful activities, but if we observe, we find that the unlawful activities related to Bitcoin have decreased over time with the starting of nontransparent cryptocurrencies. Also, users look to be unfortified because payments are always unalterable, and an inaccurate transmission cannot be scrapped, not likely credit card payments (Böhme et al. 2015).

On the positive side, the improvement of the cryptocurrency market makes it easier to access finance (Adhami et al. 2018). The starting of the blockchain technology authorized entrepreneur of businesses and their teams to elevate capital investment in cryptocurrencies and fiat currencies with the issuance of tokens digitally (Initial Coin Offerings, ICOs) and the progress of 'smart contracts' (Giudici and Adhami 2019). Tokens make it easier for buyers to utilize some products or services of the issuer, to distribute revenues, and they look the same with equity. Special crypto exchanges work as a market that is secondary for tokens. The OECD (2019) presents elementary principles and distinctive ways of an ICO. Now, this crypto asset holds more resemblance to conventional assets. Anyone would expect that the primary forecast of neoclassical finance should come true. In fact, in the current factual study regarding crypto tokens, Howell et al. (2018) demonstrated the impact of crooked data on trading of tokens; both trading volume and its liquidity are directly related to the information influx. The latter is accomplished through intended exposure of information (including their business plans and operating budget) and signaling quality.

Cryptocurrencies are a lookalike to ICO norms that offer unbiased and autonomous access to both high productivity as well as the capital, in comparison with fiat money, that allow peer to peer transactions and avoid intermediation of banks (Nakamoto 2008; Karlstrøm 2014). It is usually done through ICO. Businesses, which have narrow finance lines, gaps in the capital, and the ability to compete with specialized investors, can use it as a relevant opportunity. OECD (2019) reported that ICOs are a prospective path for low-cost finance for SMEs.

Does cryptocurrency play a role and help in the process to democratize funding? Nowadays, consultants, investors, and practitioners are widely discussing this possibility with a vast type of views. For instance, The World Economic Forum White Paper (WEF 2018) affirms how blockchain technologies and cryptocurrencies can enhance trading volume worldwide, improve higher service levels and minimal transaction fee structure. To this level, the work done by Ricci (2020) in his study considered the geographical network of Bitcoin transactions to identify prospective connections between Bitcoin exchange activities among national levels of monetary autonomy in different countries. The study showed that a high level of freedom to trade internationally guarantees low tariffs and facilitates international trade which is robustly linked to Bitcoin dispersion. The freedom to deal globally will bring growth to foreign trade through payment mechanisms by lowering transaction costs like cryptocurrencies, and lowering capital limits will boost the cryptocurrency usage for unlawful dealings like money laundering.

The recompense process for miners of cryptocurrency produces an inducement to hold computing power and increase energy consumption. For instance, Böhme et al. (2015) observed that computational efforts of miners are huge on cost, the reason being that the proof work calculations are "power-intensive, which consume more than 173 MW of electricity consecutively. From the standpoint, the amount is almost $178 Mn each year at an average of US domestic electricity prices." This significant issue was raised by Vaz and Brown (2020) who suggested that there are sustainability issues in the cryptocurrency expansion beyond possible paybacks that are bagged characteristically by a few people. Consequently, the demand for diverse and alternate institutional models is motivated largely by private money and profit motivations.

12.5 APPLICATIONS OF CRYPTOCURRENCIES

Cryptocurrency can be a great way to promote ethical business practices. The usage of blockchain technology makes it possible to track all the transactions which promote complete transparency.

12.5.1 Travel

The travel industry is one industry where cryptocurrency is being used to a large extent. Travel is the second most growing industry worldwide. Cryptocurrency can make travel much more convenient. For instance, if a person uses Bitcoins as a way of payment, a lot of excessive costs associated with the freight could be saved as cryptocurrency doesn't have ATM withdrawal costs or transaction fees that are generally associated with international credit card use. Cryptocurrency has become more popular among many websites worldwide. For instance, a US-based travel company, where you can book flights, book a cruise, and rent a car has started accepting cryptocurrencies as a means of payment since 2013. Sometimes, travelers don't use cryptocurrencies for using it as a mode of payment but they use it for exchanging their fiat currencies with digital currency.

12.5.2 Education

Educational institutions based in Germany, the US, and Cyprus are now accepting cryptocurrency-based payments. Sources state that a university in Cyprus is the first in history to accept tuition fees through BitPay. Blockchain can be of great help by facilitating a direct connection between educators and students and can also provide less costly learning platforms. Such platforms allow educators to create a different learning platform for students and earn tokens in return.

12.5.3 Banking and Financial Services

Cryptocurrencies have changed the way people conducted all banking transactions. It not only provides the services that all the traditional banks do but also, besides, it provides transparency, a more efficient way of conducting transactions and security to its users. Cryptocurrency results in either free or very low cost of transactions.

Suppose a user has to pay ₹ 100,000 to his friend in another city, and the user pays it using a credit card which will raise the transaction cost, whereas if the user had transacted via cryptocash, it would have led to very low or free charges for the transaction. Traditional banking is designed which favors the people who have sound knowledge in the field of finance and fiat currencies. Nowadays, lending has become one of the areas which are increasing at a very fast rate. People are taking fiat loans giving cryptocurrency as collateral and vice versa. On networks like Etherum, many lending services such as Maker, Compound, and Instadapp are making success in the loan business with millions of thousands of dollars locked up in their protocol. There are other services also which ask the cryptocurrency users to lock up their crypto tokens in return of an annualized rate of return.

12.5.4 Wealth Management

Management is also an application of cryptocurrency. Many investors have claimed that they have earned a handful of returns by investing in cryptocurrency. The most important application of this kind of currency is that it can be used to hedge against the dollar. A Canada-based company called SwissBorg has created tokens for its users that can be used as an investment and are tradable through which a user can earn profit just like the stock market. It lets the user manage their investments by providing them great opportunities to manage their portfolio without any restrictions.

12.5.5 Wide Usage and Easy Accessibility

You can buy anything, even a Ferrari. If a person has a deep pocket digitally, he can buy anything using cryptocurrency. There is a separate marketplace on the platform on which people can buy the most expensive things like wines, sports cars, diamonds, and many more things. People who don't want to disclose the accountability about their money and want to invest in something usually prefer to buy expensive things using cryptocurrency. With the coming up of digital era, over 2 billion people have access to internet around the globe today. Some believe that one should maximize the potential of internet and replace the conventional payment exchanges. So, they can opt-in for Bitcoin as it is easily accessible for everyone with just the ease of internet.

12.6 LEGAL ISSUES ASSOCIATED WITH CRYPTOCURRENCY

The anonymity feature of cryptocurrency trading makes it difficult for government bodies and regulatory bodies to detect suspicious activity on the platform. Some countries like Bolivia, Kyrgyzstan, and Ecuador have banned the use of cryptocurrencies. China has strictly asked its financial institutions and banks to not deal with cryptocurrencies. Terrorists can obtain funds in the form of tokens because of the deregulated feature of cryptocurrency. Terrorism financing schemes are growing each day. According to a New York-based research company, in 2019, a designated terrorist organization raised money from a website that makes Bitcoins. This terrorist organization appealed to its followers to transfer Bitcoins in their accounts and even instructed and asked them to use a public Wi-Fi so that the IP address of the sender does not get traced. Money laundering and terrorist funding have been a primary concern for the governments of many countries, and they have even asked the cryptocurrency developers to get more information about the user so that the filtering of people involved becomes easy. Facebook's cryptocurrency Libra has been stalled by the US government due to privacy issues. In the year 2019, $829 million worth of Bitcoins were used on the dark web. So, we can get an idea of how much of cryptocurrency is being used in activities like money laundering, criminal activities, and terrorist funding. All cryptocurrencies are attractive to the lawbreakers. For instance, if a person wants to exchange tokens, then he can transfer the money into another account, and nobody would be able to track the details of the transaction due to the intractability feature. The Treasury Secretary of the USA recently talked about this issue in his last two speeches and asked the cryptocurrency developers to maintain some privacy if they don't want to get involved in any legal complexities. They said if they find anyone promoting any criminal activity, strict legal action would be taken against that person. A leading sheikh of a terrorist group Syria Hayat Tahrir Anshan posted a video of him explaining how he gets money from all the sources and how easy it is to transfer Bitcoins to his account. He asked his followers to give him some money which will be used in terrorist funding.

The main question which most of the experts and analysts ask is "Should we not ban some aspects related to Bitcoin?" Technologies like private send, stealth addresses, and Kovri projects are features of Bitcoins which make cryptocurrency untraceable. Now another question arises, "Is it necessary to make these features applicable to this platform?". Imposing a ban on these technologies is in line with the Council's conclusion 2018 on how to tackle cybercrime and criminal activities on platforms.

But whatever the answer be, we should start being intelligent and a smart investor and stop being naïve. A ban would make the transactions traceable, but how to identify any breach in it and to protect ourselves from cybercrime would be solely our responsibility. It is very likely to happen that imposing a ban on cryptocurrency could be detrimental for both the investors and the concerned regulatory bodies and authorities.

In any case, imposing a ban on the specific illicit use of cryptocurrency should not be an area to look upon; in fact, there should be a ban only on some areas which pose a problem and raise questions on illegal activities. General activities like transferring money and Bitcoin should not be barred. This paper is not in favor of banning the system as a whole; rather, the regulatory body and concerned authorities should keep track of all illegal activities. The government should play a crucial role in this. As long as this platform is benefiting general investors, it should be promoted.

12.7 ENVIRONMENTAL ISSUES ASSOCIATED WITH CRYPTOCURRENCY

As already discussed, cryptocurrency runs on the technology of blockchain. From a layman's perspective, blockchain can be interpreted as a chain of blocks, wherein each block is responsible for carrying information about a cryptocurrency transaction that took place. The block generally contains information like the date, time, and amount of the transaction. Subsequently, when the next transaction takes place, the data about the same get saved in the block and the chain continues. All the blocks in the chain are distinctive and store data in the form of unique codes called 'hash'. Hash is cryptographic codes, which are securely encrypted with special algorithms and are tremendously hard to the crackdown.

Before entering the blockchain, the authenticity of the transaction needs to be verified and validated to process the transaction further, and this process is popularly known as 'mining'. Mining is not as easy as it sounds; it requires strenuous efforts from the crypto miner. The crypto miner has to solve complicated mathematical problems to decode the block and verify the information, and the transaction must contain a proof of work (POW) to be considered as valid. Under the POW model of mining, there is a competition between several crypto miners, and the first one to do so gets rewarded in cryptocurrency. The increasing daily on-chain transactions have made the verification process very complex over time; it is like tying a knot over a knot after every transaction. But the reward for mining is so alluring that it facilitates more and more miners to compete in the process, and this has given rise to a technological 'arms race'.

To win this technological 'arms race', miners explicitly update their computer processors and infrastructure for cryptocurrency mining, thereby causing a significant impact on the electricity consumption. High consumption of electricity is just one side of the problem; miners are inclined on incentivizing unsustainable energy. Therefore, the environmental impacts of cryptocurrency have been drawn in two directions; firstly, the mining process which requires a considerable volume of energy, and secondly, the crypto mining in locations that generate unsustainable energy.

The research estimated the total electricity consumption of cryptocurrency by carefully examining every stage involved in crypto mining, from the extraction of raw materials to the production and till the very recycling of the equipment. They have projected that about 99% of the environmental impact is from using the computer setup and equipment. Calculations stated that the cryptocurrency network consumed 64.15 terawatt hours of electricity and emitted about 17.3 megatons of carbon dioxide and equivalents. Most of the crypto miners tend to reside in areas like inner Mongolia and China, where most of the electricity is produced using fossil fuels, which makes the electricity cheap over there and also areas which have a cool climate, as less energy is needed to cool down the computers.

Unsustainable energy tends to contribute more to carbon footprint when compared with sustainable and recyclable energy. Cryptocurrency accounts for approximately 0.5% of the world's total electricity and exceeds the average yearly energy consumption of countries like Switzerland and the Czech Republic.

12.8 SCAMS

Cryptocurrency has always been considered a risky avenue for investors to put their hard-earned money mainly because of its volatile nature and also because of several secondary factors such as deregulated nature, fear of online theft, frauds, and no backing from the national government. Some of the scams that took place in the recent history of the cryptocurrency industry are discussed in this section.

- **PlusToken Scam**

    PlusToken scam of 2019 is the biggest scam that has taken place in the short history of cryptocurrency; it was carried out by a Ponzi scheme (a deceit scheme where the project offers a high rate of returns while bearing low risk). The scammers defrauded a sum of about $2.9 billion in cryptocurrency out of the investors by incorporating a fictitious project, based in South Korea. They said it is a crypto wallet project, and investors will get high interest on their deposits in the form of cryptocurrency. Plus Token told its investors that all the highest interest payments would be generated from the operations of the company. In the sight of their luring scheme, around 3 million investors registered for the scheme and by the end of 2019, the company exited with all the depositor's money.

    The scammers were not lucky enough, soon PlusToken was exposed as deceit. The Chinese authorities responded quickly and arrested six individuals connected to the PlusToken scheme.

- **Malware Scams**

    Malware holds a prominent name in online scams. Malware scams is the threshold for cryptocurrency scams considering its dysregulated and anonymous nature. A tech support site named Bleeping Computers has recently issued a warning to its customers about a cryptocurrency malware named cryptocurrency Clipboard Hijackers.

    Cryptocurrency Clipboard Hijackers gets into the window of the user and starts recording the clipboard for cryptocurrency transactions, and if one is pinned out, they'll immediately swap the transaction address with their address, and thus, the cryptocoin of the user will be transferred to the scammer.

- **Fake Bitcoin Exchanges – BitKRX**

    It is considered to be one of the easiest ways to scam the investors, by posing as a division of an existing recognized and respectable organization.

    BitKRX scam took place in South Korea, wherein BitKRX claimed itself to be a cryptocurrency exchange (a place where cryptocurrencies can be traded), a division of the real Korean Exchange (KRX), a subbranch of KOSDAQ. BitKRX scammed the users by calling itself a legitimate cryptocurrency exchange and ate all the crypto coins that came for exchange in BitKRX.

12.9 IMPLICATIONS OF CRYPTOCURRENCY

The exponential growth that cryptocurrency enjoyed between the years 2016 and 2018 was magnificent; cryptocurrency needed a good kick start for introducing itself to the world, and it got it. History states that whenever a new industry is born which offers something new to the customers, it tends to experience exponential growth in its initial years, but this does not imply that the industry will grow at the same pace throughout its expectancy. Cryptocurrency is still in the pipeline and is not yet established, so it is difficult to comment on its future and implications.

Cryptocurrency has opened new doors for the finance sector; it has revolutionized the whole industry with its foundation of technologies. Cryptocurrency has been a one-stop solution for all the problems that were associated with the conventional finance system. Cryptocurrency is popular in the western and middle eastern countries and is slowly moving toward the traditional market. Markets in these countries are flooded with cryptocurrency transactions, some countries like the UAE, Russia, and Sweden are also coming up with their national cryptocurrency, marking validity. But before introducing a cryptocurrency to the mainstream market, extensive research, and analysis of its repercussions on the economy at the micro and macro levels should also be studied.

Currently, countries running on a single fiat currency are still not able to sustain themselves. Multicurrency economies hold a solid financial infrastructure with several legislative and regulative norms, to manage unbalances in the economy. Also, there is no hint about how it will be adjusted and contribute to the GDP.

Cryptocurrency is still in its infancy, undergoing innovations and technological upgrades to improve and become more user-centric. Simultaneously, it is also facing price fluctuations, making it harder for people to be speculative about it. It might be possible that if cryptocurrency gets familiarized with people, there is a possibility that it will become less volatile over time. But as per the current scenario, it is hard to predict the volatility of cryptocurrency. Cryptocurrency has still not convinced economists that it will be a good store of value for the public.

If cryptocurrency starts getting circulated as a fiat currency, and all other industry products are also indexed to cryptocurrency, they will also tend to have fluctuating prices as the merchant will not know what the price of Bitcoin would be the following day. Suppose, on the 1st of April, the cost of a dining table is 0.01 Bitcoin, and the price of Bitcoin is $63, on 2 April, the price of Bitcoin becomes $80, so consequently, the price of the dining table will also rise. Even if it gets implemented, the price determination will be an issue, currently, it's being determined by the forces of demand and supply, but if it has to run as a fiat currency, it needs to be backed up by an asset.

12.10 SCOPE OF GROWTH OPPORTUNITIES FOR CRYPTOCURRENCY

The story of cryptocurrency has never been less than a thriller film, the skyrocketing demand it enjoyed in the initial half of the last lustrum, slump in the middle of it and then balancing in the lateral part. Cryptocurrency does not fail to amuse the investors that they are bidding on a rollercoaster-ridden currency. Cryptocurrency always had a gigantic reward for the people who were ready to face the risk associated with its downside, but unseeingly if you go for any financial security in the world yielding high returns, it is sure that it will have a downturn as well. Still, the scenario of cryptocurrency is far different from other financial securities as it entails some exceptional growth opportunities in the future that will transform the archaic financial system.

Cryptocurrency has got many underlying opportunities in the future. It does not require any intermediary to exchange value, which means more flexibility for trading; intermediaries, mostly banks, facilitate payments between two parties. As per the Global Findex, only 69% of the global population have a bank account (*source: The World Bank*), and on the other hand, 86.6% of the global population have internet access (*source: The International Telecommunication Union*). So, marginally, 17.7% of the global population would be able to transact value via cryptocurrency above any other mode of payment which requires an intermediary. Also, transacting via cryptocurrency is relatively easier than any other mode of payment as it's decentralized and involves less paperwork. Transfer of payments and trades in millions of dollars can be done in the blink of an eye.

In the digital era of globalization and international business, conglomerates and big corporations are expanding to every corner of the world, and even small-scale industries are not stepping back to expand their operations across borders. Producers and manufacturers are acquiring foreign capital and the latest types of machinery from international markets to operate and maximize their returns providing consumers with better quality and a high range of products, which are not accessible to them in the local markets. International payments and transfers have a necessity now. If we go by the dynamics, cryptocurrency tends to overpower the traditional financial system; the thriving demand of ecommerce in the marketplace and the conventional ways of the financial system are not substantially meeting them, they take a long time in transacting values and can only transfer money to an extent, and even if they do, it's chained with high regulations. Apart from these issues, they also charge huge amounts of fees for these types of transactions. On the contrary, cryptocurrency with its quick and easy way of transaction can transfer huge amounts of money in a fragment of seconds. In a situation, where a company operates internationally and requires emergency funds from a cross border company, the traditional financial system will fail to meet the expectations. Hence, cryptocurrency has a big opportunity to capture this platform of international payment transfer and foreign exchange.

The generation of cryptocurrency has been restricted to a certain number so that the price of the cryptocurrency cannot get inflated by generating cryptocurrencies abundantly. This is the value proposition that facilitates cryptocurrency's rarity. As per the laws of demand and supply, if the demand for a commodity is less and supply is more, it will tend to have a higher price in the market. In the same way, cryptocurrency can be placed as a commodity in the market just like gold and other commodities. The rising popularity of cryptocurrencies has been on a toll in the last decade. This popularity gave investors the confidence to accept and trade cryptocurrencies globally. Prospectively, cryptocurrency on the footprints of gold can turn out to give it the standard of the world currency, subsequently aligning all the international currencies against the price of cryptocurrency.

12.11 CONCLUSION

Recently, academic literature has been giving enormous attention to cryptocurrency, scrutinizing how cryptocurrency is expected to muddle world economies and inflating a bubble, which could be easily speculated by criminals and could lead to a huge launder of money.

Underpinning the first view, it is often discussed that cryptocurrency needs a distinct market for itself, facilitating more secured and faster payment transaction systems, disintermediating monopolies of financial institutions. Parallelly, critics have pointed out that volatile nature makes cryptocurrencies more of speculative tradable security rather than a new kind of currency, persisting in a digital space.

The truth lies in the middle of these two arguments, with cryptocurrencies executing constructive functions and computing economic quotient, still being abruptly volatile. The focus should be on the legislation of cryptocurrencies, and generally to all digital currency, to increase the trading of crypto assets on a government-recognized exchange.

The main problem that needs attention is the battle against money laundering, tax evasion financing, and terrorist financing that is happening via Bitcoins. The current legal framework is not ample to resolve this current problem prevailing in the economy. The current legal framework needs substantial changes to address these issues and to become more suitable at the global level, as crypto transactions cannot be tamed by geographic borders.

REFERENCES

Adhami, S., Giudici, G., & Martinazzi, S. (2018). Why do businesses go crypto? An empirical analysis of initial coin offerings. *Journal of Economics & Business,* 100, 64–75.

Baur, D. G., Hong, K., & Lee, A. D. (2018). Bitcoin: Medium of exchange or speculative assets? *Journal of International Financial Markets, Institutions, and Money,* 54, 177–189.

Böhme, R., Christin, N., Edelman, B., & Moore, T. (2015). Bitcoin: Economics, technology, and governance. *Journal of Economic Perspectives,* 29(2), 213–238.

Bouri, E., Molnár, P., Azzi, G., Roubaud, D., & Hagfors, L. (2017). On the hedge and safe haven properties of Bitcoin: Is it really more than a diversifier? *Finance Research Letters,* 20, 192–198.

Caballero, R. J., & Krishnamurthy, A. (2008). Collective risk management in a flight to quality episode. *The Journal of Finance,* 63(5), 2195–2230.

Corbet, S., Lucey, B., Urquhart, A., & Yarovaya, L. (2019). Cryptocurrencies as a financial asset: A systematic analysis. *International Review of Financial Analysis,* 62, 182–199.

Corbet, S., Meegan, A., Larkin, C., Lucey, B., & Yarovaya, L. (2018). Exploring the dynamic relationships between cryptocurrencies and other financial assets. *Economics Letters,* 156, 28–34.

Dow, J., & da Costa Werlang, S. R. (1992). Uncertainty aversion, risk aversion, and the optimal choice of portfolio. *Econometrica: Journal of the Econometric Society,* 60(1), 197–204.

Fantazzini, D., & Zimin, S. (2020). A multivariate approach for the joint modelling of market risk and credit risk for cryptocurrencies. *Journal of Industrial & Business Economics,* 47, 19–69.

FigàTalamanca, G., & Patacca M. (2020). Disentangling the relationship between bitcoin and market attention measures. *Journal of Industrial & Business Economics,* 47, 71–91.

Foley, S., Karlsen, J., & Putnins, T. (2019). Sex, drugs, and bitcoin: How much illegal activity is financed through cryptocurrencies? *Review of Financial Studies*, 32(5), 1798–1853.

Giudici, G., & Adhami, S. (2019). The governance of ICO projects: Assessing the impact on fundraising success. *Journal of Industrial and Business Economics*, 46(2), 283–312.

Giudici, G., & Paleari, S. (2000). The provision of finance to innovation: A survey conducted among Italian technology based small firms. *Small Business Economics*, 14(1), 37–53.

Glaser, F., Zimmermann, K., Haferkorn, M., Weber, M. C., & Siering, M. (2014). Bitcoinasset or currency? *Revealing Users' Hidden Intentions, ECIS*, April 15, Tel Aviv.

Haryanto, S., Subroto A., & Ulpah M. (2020). Disposition effect and herding behavior in the cryptocurrency market. *Journal of Industrial & Business Economics*, 47, 115–132.

Hayes, A. S. (2017). Cryptocurrency value formation: An empirical study leading to a cost of production model for valuing bitcoin. *Telematics and Informatics*, 34(7), 1308–1321.

Howell, S. T., Niessner, M., & Yermack, D. (2018). *Initial Coin Offerings: Financing Growth with Cryptocurrency Token Sales* (No. w24774). Cambridge, MA: National Bureau of Economic Research.

in India: Its challenges & potential impacts on legislation. doi:10.13140/RG.2.2.14220.36486.

Kahneman, D., & Tversky, A. (1979). Prospect theory: An analysis of decision under risk. *Econometrica*, 47(2), 263–292.

Karlstrøm, H. (2014). Do libertarians dream of electric coins? The material embeddedness of bitcoin. *Scandinavian Journal of Social Theory*, 15(1), 25–36.

In M. Diehl, B. Alexandrova-Kabadjova, R. Heuver, & S. Martínez-Jaramillo (Eds.), *Analyzing the Economics of Financial Market Infrastructures* (pp. 20–40). Hershey: IGI Global.

Nakamoto, S. (2008). Bitcoin: a peertopeer electronic cash system. Available at https://bitcoin.org/bitcoin.pdf, Accessed 13 September 2019.

OECD (2019). Initial Coin Offerings (ICOs) for SME Financing. Available at www.oecd.org/finance/initial-coin-offerings-forsme-financing.html, Accessed 15 September 2019.

In R. Rau, R. Wardrop, & L. Zingales (Eds.), *The Palgrave Handbook of Alternative Finance*. Basingstoke: Palgrave MacMillan.

In F. Xavier Olleros & M. Zhegu (Eds.), *Research Handbooks on Digital Transformations* (pp. 225–253). Cheltenham: Edward Elgar.

Ricci, P. (2020). How economic freedom reflects on the bitcoin transaction network. *Journal of Industrial & Business Economics*, 47, 133–161.

Shefrin, H., & Statman, M. (1985). The disposition to sell winners too early and ride losers too long: Theory and evidence. *The Journal of Finance*, 40(3), 777–790.

Shiller, R. J. (2003). From efficient markets theory to behavioural finance. *Journal of Economic Perspectives*, 17(1), 83–104.

Tsyvinski and Liu (2018). Risks and returns of cryptocurrency. *The Review of Financial Studies*, 34(6), 2689–2727.

Urquhart, A. (2016). The inefficiency of bitcoin. *Economics Letters*, 148(1), 80–82.

Vaz, J., & Brown, K. (2020). Sustainable development and cryptocurrencies as private money. *Journal of Industrial & Business Economics*, 47, 163–184.

WEF. (2018). *Trade Tech—A new age for trade and supply chain finance.* The World Economic Forum in collaboration with Bain & Company. World Economic Forum, January 2018. Available at https://www.weforum.org/whitepapers/trade-tech-a-new-age-for-trade-and-supply-chain-finance, Accessed 14 September 2019.

13 Models for Predicting Student Enrolment for Delhi-Based Schools

Kartik Kakani, Biswarup Choudhury, and Sugandha Aggarwal
LBSIM

CONTENTS

13.1 Introduction 221
13.2 Literature Review 223
     13.2.1 Research Gap 226
13.3 Research Objectives...226
13.4 Research Methodology 227
     13.4.1 Research Design 227
     13.4.2 Data Collection 227
     13.4.3 Sample Technique 227
     13.4.4 Sample Size...227
     13.4.5 Data Analysis..227
13.5 Results..228
13.6 Findings 232
13.7 Conclusion and Implications..234
13.8 Limitations of the Study 234
References..234

13.1 INTRODUCTION

Education plays a very crucial role in every stage of human life. It is the process of achieving knowledge, values, skills, beliefs, and moral habits, which helps a person to become a human being. Education helps us to understand the real value and the purpose of life. Education in the childhood stage, especially in the primary levels, is important as it sets up a base and helps shaping up his/her personality and the values that a person imbibes. The education which people receive during their childhood supplies the foundation of their physical, mental, emotional, intellectual, and social development. Education expands a person's vision, and it helps to create awareness among the people in various topics related to different fields. It helps them in developing a disciplined life and provides them with better-earning opportunities for the

DOI: 10.1201/9781003140474-13

future. In other words, education is like that best friend who always sticks by us through every thick and thin of life.

In the present scenario, having proper education is also one of the parameters of measuring a person's social status within the society. One who has acquired education, has been found to have a say on diverse topics related to different fields. Hence, for the overall development of the society and the country, having values that are acquired through proper education is necessary. However, one of the biggest drawbacks in our education system is that there many instances in which students are interested to study and educate themselves, but due to certain factors, they are unable to do so. This lack of availing the opportunity to educate themselves may be caused by a number of factors ranging from lack of money to pursue education on the student's part to having no proper schools to study. According to statistical data, the literacy rate of our country is 77.77%, while in Delhi, the male literacy rate stood at 93.7%, higher than the 82.4% among females (Nandini, 2020).

Thus, the topic that we want to throw light through this chapter is about the school enrolment rate and the various factors that influence the enrolment rate. Policymakers and education planners prepare rigorous plans to make education accessible to every nook and corner within the country. The success of any plan depends on the futuristic thinking, and hence, information about any gaps prevalent in the system as well as other additional requirements needed to alter the system needs to be easily available to the planners and policymakers. So, planners and policymakers need projected educational data such as size of enrolment, number of teachers, number of schools, etc. for undertaking the exercise of future planning in respect of educational development. These tasks cannot be successfully and effectively conducted unless the government has sufficient and reliable data about the enrolled children, in advance. It is, therefore, of paramount importance to know the various aspects of the size and structure of enrolment at different points of time. Thus, through this chapter, how essential elements such as playgrounds, ICT lab, library availability, number of classrooms, number of teachers, availability of toilet facilities, drinking water availability, etc. play a significant role in influencing school enrolment decisions is listed out (Figure 13.1).

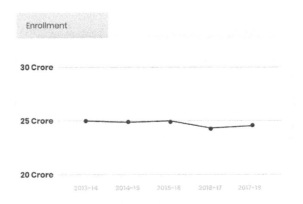

FIGURE 13.1 Student enrolment from 2013 to 2018. (Central Square Foundation, 2020.)

As shown in the above graph, the enrolment rate has declined over the years. As in 2013–2014, there were 25 crore enrolments, while in 2017–2018, there were only 24 crore enrolments (Central Square Foundation, 2020). Considering how much the world is moving forward, it is alarming how the school enrolment within our country has decreased rather than increasing in substantial numbers. So now it is high time for the education authorities to take some suitable steps so that they can arrest this decline and ensure that more children decide to enroll in schools.

13.2 LITERATURE REVIEW

Lebedinsky (2009) used panel data in this study to determine whether income inequality will lead to lower school enrolment (primary, secondary, and tertiary). The answer to this question is mixed and depends on the level of schooling and on whether they are looking at cross-country differences or variations over time. The main findings can be summarized as follows: (1) Countries with higher income inequality have lower enrolment in secondary and tertiary schools. (2) There is little evidence that changes in income inequality within a country lead to changes in secondary or tertiary enrolment. (3) In cross-country regressions, net primary enrolment is not correlated with income inequality. Gross primary enrolment is positively correlated with income inequality. (4) There is some evidence that changes in income inequality within a country might lead to lower primary school enrolment (both gross and net). (5) Other variables that were found to be statistically significant are proportion of urban population and per capita consumption expenditure. Public expenditure on schools appears to have little effect on enrolment. The only exception is tertiary enrolment, which is found to be negatively associated with government expenditure on schooling.

Afridi (2012) tried to evaluate the impact of a nationally mandated program of supplying free school meals on improving participation rates of primary school age children in a rural area of India. School panel data allow a difference-in-differences estimation strategy to address endogeneity of program placement. From the paper, the results conclude that transition of program has an enormous impact as it was seen to improve the daily participation rates of children in lower grades to a huge extent. The average monthly attendance rate of girls in grade one was more than 12% points. But coming to the attendance rate of boys, it was found that although there was a positive relation, it was not significant enough to have an enormous impact on their attendance rate, and thus, it was concluded that the overall impact that it has on the school enrolment rate was insignificant.

Mehboob et al. (2012) investigated on the major factor affecting the student enrolment. Internal (career, aptitude), external (cost, location), and social (parents, friends) factors are further classified into 11 subfactors that affect student enrolment. A self-administrative questionnaire was distributed among 251 students from higher education institutions (HEIs). The questionnaires consist of the information on 11 factors that influence enrolment. Multiple regression analysis is done on the factors. It was seen that career is the most preferred factor chosen by the student, and in addition to these facilities, financial aid and social influence are also significantly related to student choice of enrolment. Facility is the major factor impacting the student enrolment.

Hazarika and Viren (2013) examined the effect of prior participation in early childhood developmental programs, considered endogenous, upon 7–18 years olds' school enrolment in rural north India. Analyzes by age group of data from the World Bank's 1997–1998 Survey of Living Conditions in Uttar Pradesh and Bihar reveal that 7–10-year-olds, 11–14-year-olds, and even 15–18-year-olds are more likely to be enrolled in school because of having attended an early childhood developmental program when they were 0–6 years old. It is also found that this favorable effect is particularly pronounced among children from households below the poverty line.

Sahane et al. (2014) talk about the importance of data mining in machine learning and statistics. Using these data mining techniques, primary pupil enrolment was predicted using the values that were based on the data gathered from the past. Primary pupil enrolment data for the past 10 years (2004–2013) were taken from Aurangabad (M.S) district. Data cover the time series analysis of school, village, and district level. Microsoft SQL Server Data Mining Add-ins Excel 2007 tool is used for predicting pupil enrolment. From the data, it was seen that enrolment was not uniformed over the years; also for few years, a decline in student enrolment resulted in an increase in teachers. Based on the predictions, corrective measures can be taken.

Jayaraman and Simroth (2015) have done the assessment for India's midday meal scheme impact on student enrolment. The assessment is done on the government schools, and more than 420,000 schools were seen from 2002 to 2004. The variables on which the enrolment is measured are infrastructure availability (playgrounds, classrooms, toilet), staff, teaching, and learning materials (library availability, books, learning materials). Thirteen percent increase in student enrolment is because of the midday meal. On an average, 2.4 million more students have enrolled in schools because of midday meal. It was also observed that 100 Rs spent on midday meals will translate to an added year of student participation.

Haris et al. (2016) have discussed about different techniques that can be applied by HEIs to predict the student enrolment as these different techniques are important to know which technique is providing us the better result and to take the accurate decision based on that. Direct, correlative, and structural techniques are used by HEI for predicting enrolment. Further, the good prediction technology can be either qualitative, quantitative, or a mixture of both. WEKA, RapidMiner, KEEL, Orange, and Tanagra tools are used for data mining. These tools are used for understanding the activities of prediction and description.

Idrissa et al. (2017) diverted from the original way of figuring out enrolments that take place in a school by considering only educational selectivity to find out the determinants of child school enrolment in Ghana. Using data from the Ghana Living Standard Survey round 6 (GLSS 6), a three-step logical model for the determinants of secondary school enrolment is estimated and the dependence it has on the children to complete their primary school. From the findings, it was inferred that family resources such as parental education, household income, and the gender of the head of the household play a crucial role in households' child schooling decisions. Educated parents are relatively more likely to enroll their children in primary school and keep them in school until they complete primary education. As well, it is shown that educated parents do not promote a gender-biased investment in the schooling of children at the primary level. The welfare of the household does not influence children's entry into primary school, but importantly, their completion of primary school depends a lot on household welfare.

Wanjau and Muketha (2018) have distributed the questionnaire among 220 students out of which 209 responses were collected. As the need of HEIs are growing, they must be more informed and knowledgeable about the student and the choice of their selection. In this paper, the data of student's enrolment for science, technology, engineering, and mathematics using weighted ensemble classifier are used for mining. Noisy instances are removed to improve the quality of data. Here, the comparison between the single model and ensemble model is done, to show how the ensemble model has enhanced the result. The WEKA tool is used for analysis.

El-Saadani and Metwally (2019) are concerned with the impact of disability on youth' educational opportunities is scarce in Egypt. They provide a profile of youth with a disability and examine the impact of disability among youth on their school enrolment. Results revealed that two in one hundred youth live with severe disability. One-third of youth with disability have never attended school compared to less than 5% among their peers without disabilities. Disability plays a dominant role in hindering school enrolment, and it interacts with the individual's standard of living in a way that exacerbates inequity in educational opportunities (Table 13.1).

TABLE 13.1
Summarized Review of the Research Papers

Title	Author	Year	Findings
Does income inequality affect school enrolment?	Lebedinsky	2009	Countries with higher income inequality have lower enrolment in secondary and tertiary schools. Also, there is little evidence that changes in income inequality within a country lead to changes in secondary or tertiary enrolment
The impact of school meals on school participation: evidence from rural India.	Afridi	2012	The results suggest that program transition had a significant impact on improving the daily participation rates of children in lower grades
Factors influencing student's enrolment decisions in selection of HEIs.	Mehboob et al.	2012	It was observed that career is the most preferred factor chosen by the student; in addition to this facility, financial aid and social influence are also significantly related
The effect of early childhood developmental program attendance on future school enrolment in rural north India	Hazarika and Viren	2013	Living conditions in Uttar Pradesh and Bihar reveal that 7–10-year-olds, 11–14-year-olds, and even 15–18-year-olds are more likely to be enrolled in school because of having attended an early childhood developmental program when they were 0–6 years old
Prediction of primary pupil enrolment in government school using the data mining forecasting technique	Sahane et al.	2014	It was observed that enrolment was not uniformed over the years; also for few years decline in student enrolment resulted an increase in teachers

(Continued)

TABLE 13.1 (*Continued*)
Summarized Review of the Research Papers

Title	Author	Year	Findings
The impact of school lunches on primary school enrolment: evidence from India's midday meal scheme	Jayaraman and Simroth	2015	Roughly, 2.4 million additional students have enrolled in schools because of midday meal. It was also observed that 100 Rs spent on midday meals will translate to an additional year of student participation
A study on student enrolment prediction using data mining	Haris et al.	2016	The results suggest that program transition had a significant impact on improving the daily participation rates of children in lower grades
Analysis of school enrolment in Ghana: a sequential approach	Iddrisa et al.	2017	Educated parents are relatively more likely to enroll their children in primary school and keep them in school until they complete primary education. Educated parents do not promote a gender-biased investment in the schooling of children at the primary level
Improving student enrolment prediction using ensemble classifiers	Wanjau and Muketha	2018	Student background related to studies and income was considered for the enrolment. High school final grade and career flexibility came out to be the most important factor for the enrolment
Inequality of opportunity linked to disability in school enrolment among youth: evidence from Egypt	El-Saadani and Metwally	2019	One-third of youth with disability have never attended school compared to less than 5% among their peers without disabilities. Disability plays a dominant role in hindering school enrolment, and it interacts with the individual's standard of living in a way that exacerbates inequity in educational opportunities

13.2.1 Research Gap

None of the research papers that we reviewed, used multiple regression techniques such as linear regression, ridge regression, lasso regression, and ElasticNet regression and compared their results using regression analysis in python. In the research papers that we reviewed, there is no mention of the fact of basic features such as furniture available for students, number of classrooms available for students, ICT lab, laptops, Digi board, etc. that are available in the school campus and how improving them can help enhancing the enrolment rate.

13.3 RESEARCH OBJECTIVES

- To determine the crucial factors in predicting school enrolment.
- To help in the policy-making decisions of the education authorities for undertaking the exercise of their future planning.

- To compare different regression model techniques to know which one is best suited for the dataset.

13.4 RESEARCH METHODOLOGY

13.4.1 Research Design

A mixed research design, i.e., quantitative and qualitative both sets of research design have been followed while collecting the research findings in this project. Quantitative data were collected through the information contained on the government's website. Qualitative research includes the responses to the questionnaires that were filled in by the school authorities.

13.4.2 Data Collection

Data are collected from a government website **https://udiseplus.gov.in/** that contained information about how a range of factors influence school enrolment in India. Along with that, data were also collected through a questionnaire that had been handed out to the various schools upon which the research was conducted. Questionnaires consisted of ten sections; section 1 consisted of school profile information, section 2 had physical facilities and equipment information, section 3 had the information about the teaching and nonteaching staff, section 4 had new enrolment and repeater information, section 5 had information regarding incentives and facility provided to children, section 6 had annual examination result information, section 7 had board examination result information, section 8 had receipt and expenditure information, section 9 had PGI (performance grading index) information, and section 10 had school safety information. The questionnaire had both qualitative and quantitative questions about the factors affecting the school enrolment. Few of the factors are drinking water availability, availability of ramps in school, number of desktops, laptops, and tablets available in the school.

13.4.3 Sample Technique

A simple random sampling technique was used to sample 2,106 samples out of which 2,089 samples were found appropriate, by preprocessing and cleaning the data for further analysis.

13.4.4 Sample Size

A sample of 2,106 data sets was taken out of 5,703 data sets to obtain accurate results for the project and to describe and perform the appropriate quantitative statistical measures on this sample set to understand the trends better and to predict the school based on different features.

13.4.5 Data Analysis

Quantitative analysis of the secondary data is done using machine learning. First, exploratory data analysis is applied on the dataset to analyze and summarize the

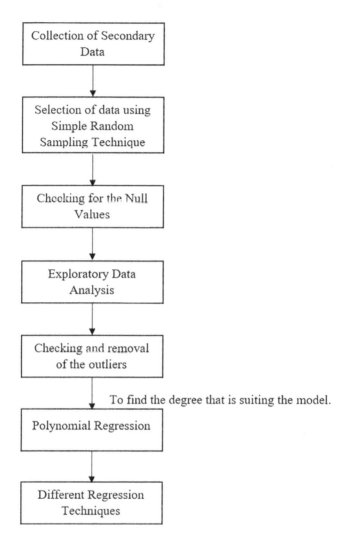

FIGURE 13.2 Flow chart describing the steps used in this chapter.

main characteristics of the data, and then, different techniques like polynomial regression, ridge regression, lasso regression, and ElasticNet regression were applied on the dataset to predict the enrolment in school. Also, a comparison of the errors between the techniques mentioned above took place, and then at the end, the best technique out of the lot is recommended to predict the school enrolment in Delhi (Figure 13.2).

13.5 RESULTS

From the dataset, data containing 2,106 sample sets are taken using the simple random sampling technique. The data consist of 40 features and 'class enrolment' as the target variable. Then, preprocessing of the dataset was done to check for the null or NA values and outliers in the data. For two of the features, scatter plots between target variable and feature are shown below. From the scatter plots (Figures 13.3 and 13.4), it can be seen that outliers in furniture are available above 5,000 and have class enrolment above 800. Also, for total classrooms, outliers lie below 20 and have class enrolment above 400.

After preprocessing the data and taking out the outliers and null values, there were 2,089 samples out of 2,106 samples, on which different regression techniques were applied to predict the school enrolment and check the error of each method.

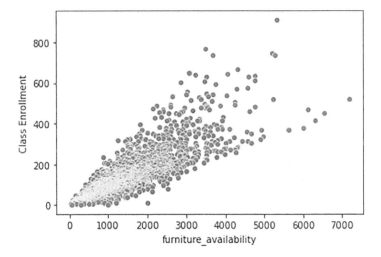

FIGURE 13.3 Scatter plot between class enrolment and furniture available for students.

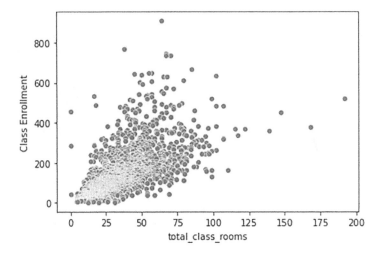

FIGURE 13.4 Scatter plot between class enrolment and total classroom for students.

Correlation has been calculated for all variables to eliminate the features having high positive/negative correlation. From the below correlation heatmap (Figure 13.5), it can be inferred that six features have high correlation, i.e., above |0.7| (urinal girls, total boys' functional toilet, urinal boys, total girls' functional toilet, classrooms in good condition, furniture availability). Out of this, the furniture availability has not been removed as it is having a high correlation with our target variable 'class enrolment'.

In feature encoding, a dummy variable is created for our categorical variable (Figure 13.6) using the pandas, and then, standard scaling is applied on all the features.

The remaining 2,089 samples are distributed into train and test sets of ratios 0.7:0.3 to apply polynomial regression. From the polynomial regression, degree 1 is coming out to be best fit, as higher degree is overfitting the model, i.e., working fine on the training set, but the error is increasing on the test set as the degree is increasing. Thus, it was decided to go through with degree 1 (Table 13.2).

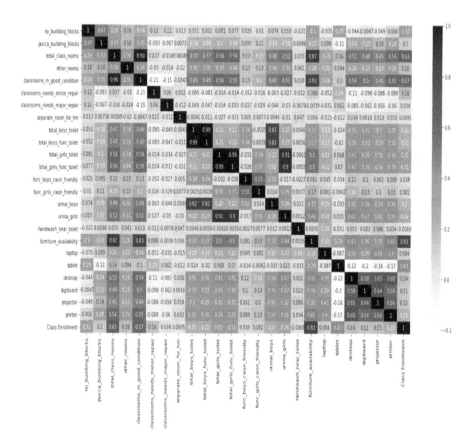

FIGURE 13.5 Heatmap showing the correlation between all variables.

Student Enrolment for Delhi-Based Schools

index	building_status	boundary_wall	drinking_water_available	drinking_water_functional	rain_water_harvesting	handwash_facility_for_meal	electricity_availability	solar_panel	library_availability	playground_available	medical_checkups	availability_ramps	internet
0	Government	Pucca	Yes	Yes	Yes	Yes	Yes	Not_func_	Yes	Yes	No	Yes	Yes
1	Government	Pucca	Yes	Yes	Yes	Yes	Yes	No	Yes	Yes	No	Yes	Yes
2	Government	Pucca	Yes	Yes	No	Yes	Yes	No	Yes	Yes	No	Yes	Yes
3	Government	Pucca	Yes	Yes	No	Yes	Yes	No	Yes	Yes	Yes	Yes	Yes
4	Government	Pucca	Yes	Yes	Yes	Yes	Yes	No	Yes	Yes	No	Yes	Yes
5	Government	Pucca	Yes	Yes	Yes	Yes	Yes	No	Yes	Yes	No	Yes	Yes
6	Government_S_	Pucca	Yes	Yes	No	Yes	Yes	Not_func_	Yes	Yes	No	Yes	Yes
7	Government	Pucca	Yes	Yes	Yes	Yes	Yes	No	Yes	Yes	No	Yes	Yes
8	Government	Pucca	Yes	Yes	No	Yes	Yes	No	Yes	Yes	No	Yes	Yes
9	Government	Barbed_wire_	Yes	Yes	Yes	Yes	Yes	No	Yes	Yes	Yes	Yes	Yes
10	Government_S_	Pucca	Yes	Yes	No	Yes	Yes	Yes	Yes	Yes	No	Yes	Yes
11	Government	Pucca	Yes	Yes	Yes	Yes	Yes	No	Yes	Yes	No	Yes	Yes
12	Government	Pucca	Yes	Yes	No	Yes	Yes	No	Yes	Yes	Yes	Yes	Yes
13	Government	Pucca_Broken	Yes	Yes	No	Yes	Yes	No	Yes	Yes	Yes	Yes	Yes
14	Government	Pucca	Yes	Yes	Yes	Yes	Yes	No	Yes	Yes	Yes	Yes	Yes
15	Government	Pucca	Yes	Yes	Yes	Yes	Yes	No	Yes	Yes	Yes	Yes	Yes
16	Government	Pucca	Yes	Yes	No	Yes	Yes	Not_func_	Yes	Yes	No	Yes	Yes
17	Government_S_	Barbed_wire_	Yes	Yes	No	Yes	Yes	No	Yes	Yes	Yes	Yes	Yes
18	Government	Pucca	Yes	Yes	Yes	Yes	Yes	No	Yes	Yes	Yes	Yes	Yes
19	Government	Pucca	Yes	Yes	Yes	Yes	Yes	No	Yes	No	Yes	Yes	Yes
20	Government	Pucca_Broken	Yes	Yes	Yes	Yes	Yes	No	Yes	Yes	Yes	Yes	Yes
21	Government	Pucca	Yes	Yes	Yes	Yes	Yes	No	Yes	Yes	Yes	Yes	Yes
22	Government	Pucca	Yes	Yes	Yes	Yes	Yes	No	Yes	Yes	Yes	Yes	Yes
23	Government	Pucca	Yes	Yes	No	Yes	Yes	No	Yes	Yes	No	Yes	Yes
24	Government	Pucca	Yes	Yes	No	Yes	Yes	No	Yes	Yes	No	Yes	Yes

FIGURE 13.6 Categorical features in the dataset.

TABLE 13.2
Comparing the Train and Test Set Error for Polynomial Regression

Degree	Train Set Error (RMSE)	Test Set Error (RMSE)
1	1.290	1.357
2	0.872	$26.28 * (10^{11})$
3	0.014	$42.158 * (10^{10})$
4	0.014	$60.10 * (10^{10})$

13.6 FINDINGS

From the model, it is found that ElasticNet regression is the best model to predict the class enrolment as it has the lowest root mean squared error (1.347) and the best R square (71.2%). Out of 40 features, the coefficient of 11 of them came out to be 0. For us, l1_ratio came as 1 which is L1 penalty. As the value is 1, it means it is much closer to the lasso regression. On the other hand, lasso regression is a much simpler model to predict in comparison to other models as it has only four features with nonzero coefficient, and the remaining 36 features have 0 coefficient.

Here from the test data set, i.e., 626 samples, the value for class enrolment is predicted using linear regression, ridge regression, lasso regression, and ElasticNet regression (Figure 13.7, Table 13.3).

In all the regression models, furniture availability, i.e., the number of students for whom furniture is available in a school, is the most important as it has the highest value of coefficient in comparison to others (Table 13.4).

It is also evident from the correlation between the target variable and features as furniture availability has the highest correlation (Figure 13.8).

Index	Actual	Linear_Regression_Predicted	Ridge_Regression_Predicted	Lasso_Regression_Predicted	ElasticNet_Regression
0	278	265	265	226	259
1	210	230	230	202	233
2	32	59	59	68.9	61.4
3	100	112	113	102	110
4	48	59.9	59.9	66	57.9
5	296	215	215	205	216
6	87	53.1	53.1	68	54.4
7	51	52	51.9	59.5	51.8
8	58	75.4	75.4	81	76.7
9	340	307	307	311	310
10	276	355	355	309	341

FIGURE 13.7 Comparison of actual, and predicted value using different models.

TABLE 13.3
Comparing the RMSE and R^2 for Different Regression Models

Regression	Root Mean Squared Error (RMSE)	R Square (R^2)
Linear regression	1.357	0.708
Ridge regression	1.358	0.708
Lasso regression	1.419	0.681
ElasticNet regression	1.347	0.712

TABLE 13.4
Highest Coefficient for Each Regression Models

Regression Model	Coefficient for Furniture Available
Linear regression	2.19
Ridge regression	2.18
Lasso regression	1.85
ElasticNet regression	2.15

```
Class Enrollment                 1.000000
furniture_availability           0.830063
total_class_rooms                0.633023
classrooms_in_good_condition     0.572341
tablet                           0.428118
no_building_blocks               0.321780
total_girls_toilet               0.309971
total_girls_func_toilet          0.307922
urinal_girls                     0.236652
total_boys_toilet                0.213265
total_boys_func_toilet           0.207162
urinal_boys                      0.198740
printer                          0.191590
classrooms_needs_minor_repair    0.159829
other_rooms                      0.156489
desktop                          0.141435
projector                        0.127406
digiboard                        0.112672
pucca_building_blocks            0.104156
laptop                           0.084175
func_girls_cwsn_friendly         0.081626
func_boys_cwsn_friendly          0.037551
classrooms_needs_major_repair    0.033785
separate_room_for_hm             0.009479
handwash_near_toilet            -0.006811
```

FIGURE 13.8 Correlation between features and dependent variable (class enrolment).

13.7 CONCLUSION AND IMPLICATIONS

The condition of school facilities in India is not up to the mark when compared to other developed countries, and especially, it is worse in rural India. Thus, our study focuses on finding out the facilities which affect the school enrolment in different schools.

From the above results, it can be inferred that furniture availability is one of the most important key factors when it comes to taking a decision of whether to enroll in a school or not. It is closely followed by total classrooms and whether the classrooms that are available are in a good condition or not. The least important factor when it comes to deciding upon school enrolment is the handwash near toilet facility, i.e., whether handwashes or soaps are near the toilet of the school. So, based on these results and findings from our machine learning techniques, it is suggested that the education authorities should give more impetus in making sure that the availability of proper furniture facilities for the students is considered within the school campus, along with having an adequate number of classrooms in a good condition.

13.8 LIMITATIONS OF THE STUDY

- The study is limited to only schools that are in Delhi.
- Difficulties contributed to the questionnaires as the questionnaires were already set up.
- Personal interaction was not possible with the school authorities.

REFERENCES

Afridi, F. (2012). The impact of school meals on school participation—Evidence from rural India. *Journal of Development Studies, 47*(11), 1636–1656.

Central Square Foundation. (2020). Retrieved from https://tinyurl.com/93w6ta2v.

El-Saadani, S., & Metwally, S. (2019). Inequality of opportunity linked to disability in school enrolment among youth—Evidence from Egypt. *International Journal of Educational Development, 67,* 73–84.

Haris, N. A., Abdullah, M., Hasim, N., & Rahman, F. (2016). A study on student's enrolment prediction using data mining. *10th International Conference on Ubiquitous Information Management and Communication 6,* 2016, 1–5. .

Hazarika, G., & Viren, V. (2013). The effect of early childhood developmental program attendance on future school enrollment in rural North India. *Economics of Education Review, 34,* 146–161.

Iddrisa, A. M., Danquah, M., & Quartey, P. (2017). Analysis of school enrolment in Ghana—A sequential approach. *Review of Development Economics, 21*(4), 1158–1177.

Jayaraman, R., & Simroth, D. (2015). The impact of school lunches on primary school enrolment—Evidence from India's Midday Meal Scheme. *The Scandinavian Journal of Economics, 117*(4), 1176–1203. doi:10.1111/sjoe.12116.

Lebedinsky, A. G. (2009). Does income inequality affect school enrolment? *Journal of Applied Economics & Policy, 28*(1): 69–100.

Mehboob, F., Shah, S. M., & Bhutto, N. A. (2012). Factors influencing student's enrolment decisions in selection of higher education institutions (HEI'S). *Interdisciplinary Journal of Contemporary Research in Business, 4*(5), 558–568.

Nandini. (2020). *International Literacy Day 2020—Kerala most literate state in India, check rank-wise list.* Retrieved from Hindustan Times. https://tinyurl.com/3sducdtj.

Sahane, M., Sirsat, S., Khan, R., & Aglave, B. (2014). Prediction of primary pupil enrolment in government school using data mining forecasting technique. *International Journal of Advanced Research in Computer Science and Software Engineering, 4*(9).

Wanjau, S. K., & Muketha, G. M. (2018). Improving student enrolment prediction using ensemble classifiers. *International Journal of Computer Applications Technology and Research, 7*(3), 122–128.

14 Analyzing the Functionality and Efficient Operability of the Youth During COVID 19

Megha Mishra, Reema Thareja, and Vidushi Singla
University of Delhi

CONTENTS

14.1 Introduction 237
14.2 Literature Review 238
14.3 Technique..239
14.4 Key Terminology..239
14.5 Results and Findings...241
14.6 Conclusions...252
14.7 Future Scope 252
References..252

14.1 INTRODUCTION

The unforeseen challenges caused by the COVID 19 pandemic have significant toll on people. Many have been questioning the appropriateness of the measures such as lockdown and social distancing taken by the government. People have expressed concerns all over the internet that the government is not doing enough to control the crisis and had these steps been taken early, the condition would not have been so worse. Due to COVID19 crisis, uncertainty and an enormous amount of change in our lifestyle have affected and worsened our physical as well as mental health. Students are facing problems due to online education system, lack of jobs, and uncertainty about their future. Some people suffered as they were even deprived of essential items like milk, medicines, vegetables, etc. during the lockdown. Most people faced huge losses even after the lockdown was uplifted.

Therefore, we assessed various problems such as health issues, education, etc. faced by people and analyzed them using machine learning algorithms such as KNN, SVM, logistic, and decision tree. The accuracy of each model was calculated, and a confusion matrix was formulated to identify the views and concerns of people.

We also gathered data using Twitter API to know about the problems, such as ban on international travel which affected students and business personals, faced by people all over the world. We generated word clouds, unigrams, and trigrams for even better understanding and then compared them with the problems and situations in India.

We deduced that despite the challenges, people wanted to learn and find ways to advance their education. Schools and colleges have moved to an online learning environment. Students and parents began to use a range of educational technology resources to support them with their studies. New tools and technologies started emerging and were used extensively and adapted by people. According to business standard.com, Indian edtech startups saw a total investment of $2.22 billion in 2020 as compared to $553 million in 2019. With advancing technologies, education technologies have become a change maker in India. The increase in investments across edtech analyzes what future innovation holds in the next decade.

The paper is organized into seven sections. Section 14.1 gives a brief introduction to the research problem and the methodological approach used. Section 14.2 provides a literary overview of how the current situation is unfavorable for people and is affecting their health. Section 14.3 specifies the techniques applied while processing the data received from the survey. Section 14.4 gives the key techniques used during the research. Section 14.5 describes the processed results using K-NN, confusion matrix, decision tree, logistic regression, and SVM model. Sections 14.6, 14.7 and the last part highlight the conclusion, future scope, and references, respectively.

14.2 LITERATURE REVIEW

Corona virus outbreak (COVID-19) has resulted in a socio-economic crisis and deep psychological distress among people worldwide. The government has taken many restrictive steps to control the spread of this disease, but these restrictions have impacted the mental health of people. Social distancing and lockdown measures have been carried out in almost every country. Additionally, most families are facing financial problems. Millions of people have either lost their jobs or are facing salary cuts which are further impacting their mental health. A recent report by the Centre for Monitoring Indian Economy Pvt Ltd shows that the unemployment rate in India has increased from 6.7% in mid-March to 23.48% and 23.52% in April and May, respectively. The unemployment rate in India has risen to 8.35% in August 2020, and many people in the private as well as government sectors are facing salary cuts. Not just in India, rampant job losses have impacted people in many other countries too.

Students already suffered much psychological stress and fear of negative academic consequences under normal conditions. As institutes have shifted to online learning methods, they face even more stress due to increased pressures to learn independently and uncertainty about their future. Further, in a developing country like India, every student does not have access to the internet and electronic devices to join their online classes. The world lost nearly 400 million full-time jobs in the year's second quarter (April–June 2020) due to the novel corona virus disease (COVID-19) pandemic, said the International Labour Organization (ILO) July 2. A recent survey by YoungMinds reported that 80% of young respondents agreed that their mental health has worsened during the corona virus outbreak. [1]

All over the country, mental health experts are also dealing with an increasing number of patients, particularly youths, showing suicidal tendencies because of a sense of uncertainty brought by the pandemic. People are mainly affected due to factors such as job loss or fear of losing a job, financial insecurity, stress, and loneliness. Data showed that the numbers of people seeking help from therapists are highest in the age group 25 to 40 followed by those aged between the groups 18 and 25.

The issue here is not only of mental health but also that people are also facing other problems such as deprivation of essential commodities, an increasing number of family issues such as divorces, quarrels, and lack of personal space during the lockdown.

But amidst all this, there is also a positive change adapted by people in their lifestyles. People have reduced the consumption of junk food and have also started taking a healthy diet.

14.3 TECHNIQUE

In this section, we have discussed various techniques that we have applied to learn how COVID-19 has impacted the life of our respondents. But before discussing the techniques, we have shortlisted the steps involved in conducting our research. [2]

> Step 1: An extensive research was done to finalize questions to be asked in the questionnaire.
> Step 2: 170 people in the age group of 17–30 years were held, and they were requested to fill the questionnaire prepared.
> Step 3: Data collected were organized in an excel sheet, and the sheet was loaded in an R for analysis.
> Step 4: A Twitter developer account was created, and after getting access to it, key terms were chosen to gather tweets regarding the effect of COVID 19 on people.
> Step 5: The input data were checked and cleaned to remove incorrect or redundant data.
> Step 6: Various data analysis tools were used to understand, interpret, analyze, and draw conclusions from the data.
> Step 7: Graphs were drawn to compare the effect on world and India.

Various machine learning algorithms were applied to thoroughly analyze the data.

14.4 KEY TERMINOLOGY

Logistic Regression: Logistic regression is a classification algorithm which makes use of a function to model a relation between categorical dependent and independent variable. The independent variable can be continuous, nominal, or ordinal. The value of the output ranges between 0 and 1.

$$0 \leq h_\theta(x) \leq 1$$

Logistic function (sigmoid function) is an S-shaped curve that takes any real number and maps it into a value between 0 and 1 [3].

$$f(x) = 1/1 + e^{\wedge}X \qquad (14.1)$$

K-Nearest Neighbors: It is a supervised learning algorithm which can be used for both regression as well as classification problems. It uses feature similarity to predict new values of data points. The value is assigned on the basis of how closely it matches the points in the training set [4].

There is no assumption about data points in this algorithm. In KNN, we use simple Euclidean distance formula.

$$d(p,q) = d(p,q) = \sqrt{(q_1 - p_1)^2 + (q_2 - p_2)^2 + \ldots + (q_n - p_n)^2}$$

$$= \sqrt{\sum_{i=1}^{n} (q_i - p_i)^2}$$

Accuracy – Accuracy is the ratio between the number of correct predictions to the total number of predictions. It has relatively higher accuracy.

$$\text{Accuracy} = \text{No of correct predictions/Total number of predictions} \qquad (14.2)$$

Confusion Matrix: It is a way to evaluate the performance of a classifier on a set of test data for which true values are known. The matrix is divided into predicted values and actual values. We calculate different parameters such as accuracy, precision, etc. using confusion matrix.

Decision Tree: It is a kind of supervised learning algorithm which can be used to solve both regression and classification model. The internal nodes of the tree represent features of the dataset, the leaf node represents the outcome, and branches represent the rules. It is a way of getting all the solutions to a problem based on a condition. The tree is further split into subtrees based on the answer (Yes/No) [5]. It is very easy to understand since it shows a tree-like structure. To predict the class of dataset, it begins with the root node and compares the value of root with record and then on the basis of comparison consequently follows the branch and jump to next node [6].

SVM: Support vector machines fall under the category of supervised learning algorithm used for classification and regression. It is used to find a hyper plane in an N-dimensional space that classifies data points. It divides the dataset into classes to find a maximum marginal hyper plane. For this, it first generates the hyper planes iteratively that segregates classes in best way and then chooses the best hyper plane that classifies correctly. SVM can be efficiently used to perform a nonlinear classification as well.

Naïve Bayes: Naïve Bayes is a classification technique based on probability's Bayes' theorem presenting an assumption of independent predictors. It assumes that the presence of a particular feature in a class which is unrelated to the presence of any other feature. Bayes' theorem presents a way to calculate posterior probability $P(c|x)$ from $P(A)$, $P(B)$, and $P(B|A)$ [7].

$$P(A|B) = \frac{P(B|A)P(A)}{P(B)} \tag{14.3}$$

where

$P(A|B)$: posterior probability of the class (c, target) given predictor (x, attributes).
$P(A)$: the prior probability of class.
$P(B|A)$: likelihood which is the probability of predictor given class.
$P(B)$: prior probability of predictor.

14.5 RESULTS AND FINDINGS

Tables were created to represent various attributes of the dataset to find out the relationship and effect of one attribute on the other. They were used to describe two variables (categorical) together.

In Figure 14.1, a two-way table was created to study the relationship between the impacts of COVID-19 on the personal life of people in India and where have they utilized their time mostly (hobby, taken up an online course, with family). It is seen that 92 people who have spent their most time with family are facing anxiety issues due to staying at home.

In Figure 14.2, the bar graph was created to see the ratings given by people to the government in the favor of student's education during COVID. It was observed that maximum people have given an average rating of 7–9 out of 10 to the government.

In Figure 14.3, the boxplot was created to see the effect of COVID 19 on the mental health ("good", "very bad", and "pretty much same") of people in India. We can see that the maximum number for people having very bad mental health is high.

	Hobby	Taken up an online course	With family
Anxiety due to staying at home	26	14	92
Brought closer to family	79	46	32
Pretty much same	26	14	92
Strained relationships	79	46	32
	Work/Study		
Anxiety due to staying at home	38		
Brought closer to family	14		
Pretty much same	38		
Strained relationships	14		

FIGURE 14.1 Table between impact of COVID and where most of the time is spent.

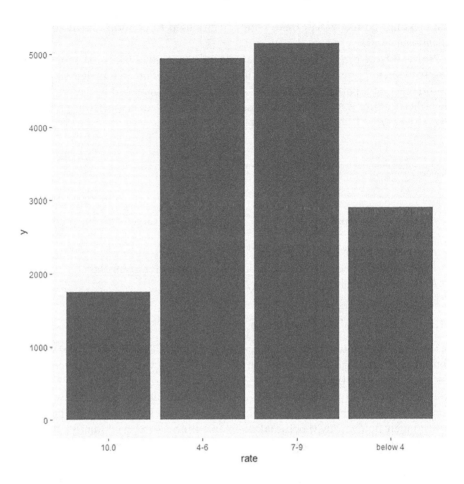

FIGURE 14.2 Bar graph to see ratings given to the government.

A word cloud graphically shows the occurrence of various words in the tweet corpus [8]. Here in Figure 14.4, we observe that words like mental health, stress, COVID, pandemic, pressure, anxiety, uncertainty, and depression are greater in size depicting how frequently these words were used to talk about mental stress in COVID. In our survey also in India, we can see in Figure 14.4 that a greater number of people have very bad mental health, whereas Figure 14.5 talks about the use of words such as job, loss, need, new, COVID, challenges, and low wage to depict that a greater number of people are suffering from job loss in the world during COVID which also concludes the reason for average rating (7–9) given to government in favor of students' education in India. We have also used a documentation matrix to depict the same.

A document-term matrix [9] is a mathematical matrix that arranges words according to their frequencies in the recovered tweets. The rows depict the documents of the tweets which are collected, and the columns correspond to the number of terms. Here in Figure 14.6, we observe the occurrence of terms like predictions, COVID, job, loss, many, and food in different tweet documents.

Functionality of the Youth During COVID-19 243

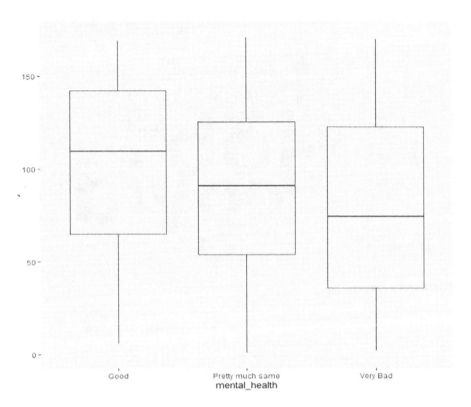

FIGURE 14.3 Boxplot defining the effect on mental health.

FIGURE 14.4 Word cloud showing mental stress.

FIGURE 14.5 Word cloud showing job loss.

```
> inspect(mat)
<<TermDocumentMatrix (terms: 743, documents: 100)>>
Non-/sparse entries: 1335/72965
Sparsity           : 98%
Maximal term length: 18
Weighting          : term frequency (tf)
Sample             :
          Docs
Terms      2 37 4 41 60 64 65 66 87 91
  ...      1  0 1  0  1  1  1  1  1  0
  americans 0  0 0  0  0  0  0  0  0  0
  covid    1  1 1  1  1  1  1  1  1  1
  food     0  0 0  0  2  2  2  2  0  0
  job      0  1 0  1  1  1  1  1  1  0
  loss     0  1 0  1  1  1  1  1  1  0
  many     0  0 0  0  1  1  1  1  0  0
  people   1  0 1  0  0  0  0  0  0  0
  the      0  0 0  0  0  0  0  0  1  0
  trump    0  0 0  0  0  0  0  0  0  0
```

FIGURE 14.6 Document-term matrix.

Pie charts were drawn to convey the information more easily. It has helped to show the percentage of data selected in each category.

In Figure 14.7, a pie chart was used to analyze the percentage of people in India who have switched over to our heritage of using Ayurvedic medicines or home remedies to boost their immunity during COVID-19. It was observed that a large percentage of people were either taking Ayurvedic medicines (44%) or were using home remedies (15%).

In Figure 14.8, a pie chart was drawn to find the percentage of people giving different ratings to government's steps in prevention against COVID 19. The pie chart

Functionality of the Youth During COVID-19   245

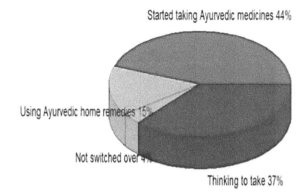

FIGURE 14.7 Pie chart depicting switch over to Ayurvedic medicines.

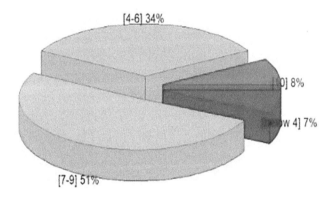

FIGURE 14.8 Pie chart depicting ratings given to the government.

showed that only a small percentage of people (8%) gave 10/10 rating to government, while majority (51%) gave an average rating (7–9/10).

In Figure 14.9, a graph is plotted against the different modes of education (online, offline), effect of online classes on education (adversely affected, found it better than online classes, did not affect much, found it productive), and effect of COVID 19 on the future plans (uncertain about future, confident about future, temporarily affected, did not affect much) to describe how the future plans of students are affected due to online classes.

Unigram: The graph in Figure 14.10 displays the text unigrams of the Twitter data regarding the job loss of people in COVID 19 throughout the world for the key terms which are mostly used to depict the terms which are least used. The y-axis gives the number of key terms, while the x-axis gives the key terms. Words such as COVID, online, difficult, new, and job are the most used words showing how frequently these words were used not only in India but also around the globe to discuss about online education during COVID.

In Figure 14.11, a dot plot was plotted between the effect of COVID 19 on personal life (anxiety issues, brought closer to family, same or strained relationships); the no.

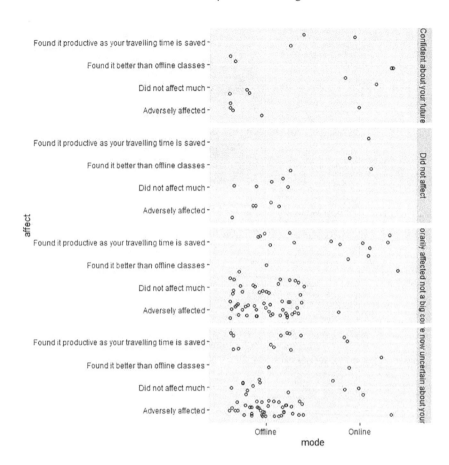

FIGURE 14.9 Scatter plot of effect of online classes on education.

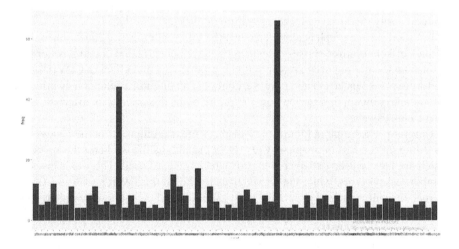

FIGURE 14.10 Unigram showing job loss throughout the world during COVID.

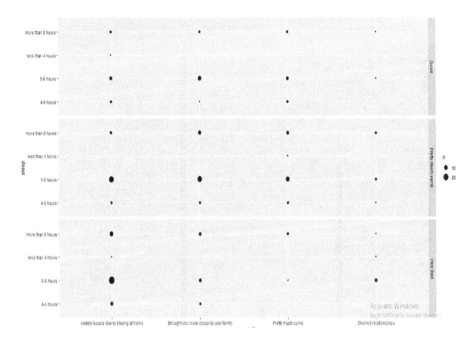

FIGURE 14.11 Dot plot depicting various factors affecting mental health.

of hours they sleep (4–6, 6–8, less than 4, more than 8) has affected the mental health of people. It describes the impact of the COVID-19 situation on personal life and sleep patterns of people and consequently their effect on mental health. In the graph, we can see that those who got anxiety issues by staying in the house claim to have a very bad mental health and sleeps for 6–8 hours (Figure 14.12).

The graph displays the text trigrams of the Twitter data regarding the consequences of the mental stress of people during COVID-19 worldwide for the key terms taken three at a time. The y-axis gives the count of key terms, while the x-axis gives the phrases of the key terms (Figure 14.13).

The logistic regression model resulted in low accuracy (78%). We applied the logistic regression model to calculate which mode of education is efficient (online, offline) considering it has a linear relationship with the effect of online classes on education (adversely affected, not affected much, found it better than offline classes, did not affect much), problems faced during online classes (missed classes, nonavailability of phones and laptops, distracted in online classes, not faced any problems as such), and the effect of COVID on future plans. For calculating accuracy, it finds the smallest possible deviance between the observed and predicted values and gets to the best fit. The accuracy was low, so we implemented an SVM model on the same dependent and independent variables

The SVM model was used as a classification technique to analyze data [10]. Confusion matrix was created to find out the accuracy of the model. Figure 14.14 shows the confusion matrix created. We converted the values of our columns to

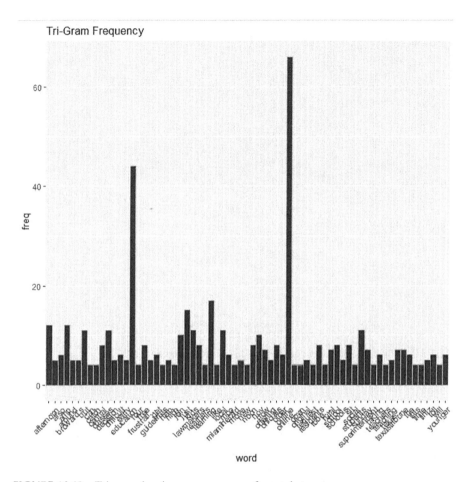

FIGURE 14.12 Trigram showing consequences of mental stress.

```
Call:
glm(formula = mode ~ affect + problems + future_plans, family = binomial(link = "lo(
    data = train)

Deviance Residuals:
    Min       1Q   Median       3Q      Max
-1.1078  -0.5283  -0.3248  -0.2884   2.5301

Coefficients:
              Estimate Std. Error z value Pr(>|z|)
(Intercept)    -3.1590     0.7188  -4.395 1.11e-05 ***
affect          1.2602     0.7217   1.746   0.0808 .
problems        1.4893     0.7204   2.067   0.0387 *
future_plans    0.2433     0.6306   0.386   0.6996
---
```

FIGURE 14.13 Logistic model.

Functionality of the Youth During COVID-19

```
Confusion Matrix and Statistics

test_pred  0  1
        0 25  9
        1  0  0

              Accuracy : 0.7353
                95% CI : (0.5564, 0.8712)
   No Information Rate : 0.7353
   P-Value [Acc > NIR] : 0.588448

                 Kappa : 0
```

FIGURE 14.14 Confusion matrix using SVM model.

```
Total Observations in Table:  43

                 | pred
prc_test_labels  |       0   |       1   | Row Total |
-----------------|-----------|-----------|-----------|
             0   |      33   |       0   |      33   |
                 |   1.000   |   0.000   |   0.767   |
                 |   0.825   |   0.000   |           |
                 |   0.767   |   0.000   |           |
-----------------|-----------|-----------|-----------|
             1   |       7   |       3   |      10   |
                 |   0.700   |   0.300   |   0.233   |
                 |   0.175   |   1.000   |           |
                 |   0.163   |   0.070   |           |
-----------------|-----------|-----------|-----------|
   Column Total  |      40   |       3   |      43   |
                 |   0.930   |   0.070   |           |
-----------------|-----------|-----------|-----------|
```

FIGURE 14.15 Confusion matrix using KNN model.

binary categorical variables (0 for offline classes and 1 for online classes). The accuracy was calculated using confusion matrix which came out to be 73.53%.

Finally, the KNN model resulted in even higher accuracy. KNN uses Euclidean distance to find the nearest value of k in the data set. Cross tables were used to check the accuracy of the two values. For this, we converted our values to binary categorical values where 1 is for online classes, and 0 is for offline classes. In Figure 14.15, it is seen that 33 out of 43 observations are True Negative and 3 are accurately predicted as True Positive. The overall accuracy of the model was 83.7%.

The logistic regression model was applied on the effect of COVID-19 on mental health considering it has a linear relationship with the number of hours people sleep and their personal life. We converted values in our column to binary categorical value where 0 represents "very bad" mental health condition, and 1 represents "good" or "pretty much same". The accuracy was 75.98%. The KNN (accuracy – 100%) model was used on the same set of columns to increase the accuracy of the model. It is a supervised learning algorithm which makes use of Euclidean distance to find nearest k in the dataset where k specifies the number of neighbors in the dataset. Accuracy is calculated using confusion matrix where 0 represents "very bad" mental health, and 1 represents "good" or "pretty much same" mental health. In Figure 14.16, it is seen that 31 are True Negative, and 21 are True Positive.

The decision tree is a powerful tool for prediction and classification. It is a flowchart which represents choices and their results in the form of a tree.[11] The data set is split, and the model is created using training data. Here in Figure 14.17, the value of c, i.e., whether personal life is badly affected or not during COVID, is compared at every iteration, and data are split accordingly. In the second node, the value of a, i.e., whether a person sleeps optimum amount of time or not is compared to generate a subnode. From the tree, we can conclude that out of 48% people whose mental health is affected due to problems in personal life, 10% are not taking optimum amount of sleep. A good clean split will create two nodes in which both have all case outcomes close to the average outcome of all cases at that node (Figure 14.8).

Naïve Bayes Classifier: Its output is both apriori and conditional probabilities for each class in the training set considering every factor attribute and class combination. Here, we calculate how frequently some particular evidence (frequcncy of the key term, i.e., mental stress during COVID 19 from the Twitter data) is observed, given a

```
Total Observations in Table:   52


                  | pred
 prc_test_labels  |           0 |           1 | Row Total |
------------------|-------------|-------------|-----------|
                0 |          31 |           0 |        31 |
                  |       1.000 |       0.000 |     0.596 |
                  |       1.000 |       0.000 |           |
                  |       0.596 |       0.000 |           |
------------------|-------------|-------------|-----------|
                1 |           0 |          21 |        21 |
                  |       0.000 |       1.000 |     0.404 |
                  |       0.000 |       1.000 |           |
                  |       0.000 |       0.404 |           |
------------------|-------------|-------------|-----------|
     Column Total |          31 |          21 |        52 |
                  |       0.596 |       0.404 |           |
------------------|-------------|-------------|-----------|
```

FIGURE 14.16 Confusion matrix using KNN model.

Functionality of the Youth During COVID-19

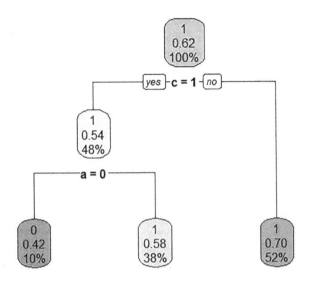

FIGURE 14.17 Decision tree.

```
> classifier

Naive Bayes Classifier for Discrete Predictors

Call:
naiveBayes.default(x = X, y = Y, laplace = laplace)

A-priori probabilities:
Y
          4           5           6           7           8           9
0.219696970 0.208333333 0.136363636 0.064393939 0.053030303 0.022727273
         10          11          12          13          14          15
0.022727273 0.022727273 0.022727273 0.018939394 0.003787879 0.022727273
         18          19          21          22          23          24
0.007575758 0.007575758 0.011363636 0.003787879 0.003787879 0.003787879
         28          31          33          39          40          41
0.003787879 0.003787879 0.003787879 0.003787879 0.003787879 0.003787879
         44          49          51          53          54          55
0.003787879 0.015151515 0.003787879 0.022727273 0.007575758 0.003787879
         58          59          61          62          74          93
0.003787879 0.003787879 0.003787879 0.003787879 0.003787879 0.003787879
         94          95         113         157         191         195
0.018939394 0.003787879 0.003787879 0.003787879 0.003787879 0.003787879
        314
0.003787879

Conditional probabilities:
    training1$freq
Y         [,1]      [,2]
   4  6.396552 0.5600644
   5  8.054545 0.7307576
```

FIGURE 14.18 Apriori algorithm.

known outcome. Conditional probability was calculated for each key term considering that the frequency of the other subject exists. Apriori probability was calculated by dividing the number of desirable outcomes by the total number of outcomes.

14.6 CONCLUSIONS

After doing extensive research on our data, we can see that most of the respondents have deteriorated their mental health during the time of pandemic and believe that government could have implemented better steps to prevent the spread of diseases. It was also observed that people throughout the world are suffering from anxiety and depression and have lost their jobs. We came to know about the problems respondents are facing during this pandemic.

These data have helped us in analyzing the situations during COVID-19 and its effect on people in India as well as throughout the world. It is seen how people in India are gradually shifting to Ayurvedic medicine. It has also helped us to analyze the various problems students are facing and how online education is still not a feasible option for many students.

The proposed technique shows us how badly the mental health of people is affected due to the changes in their personal life during COVID and the number of hours they sleep. The accuracy came out to be 100%. The technique was also applied to show a better model of education. It was observed that students faced many problems during the online mode of education like connectivity, lack of gadgets, etc. Therefore, the model when implemented helped to know about the problems students are facing in their online classes.

The paper clearly shows how the changes in the life of people during COVID are affecting them and how people have reacted to these changes and different problems faced by people like negligence of their health problems, financial stress, improper methods of learning which affected their education, etc. By mapping these patterns in their lifestyles, people can develop better habits, change their sleep patterns, or seek some medical advice.

14.7 FUTURE SCOPE

In the future, we would like to extend our project and include some more parameters to assess the situation. Since these data were collected in early August, data can be collected from December to January and assessed using various techniques. The results of these two data can be compared, and the current situation can be evaluated to know how fast things are improving or deteriorating.

REFERENCES

1. https://www.business-standard.com/article/education/indian-edtech-startups-see-investment-of-2-22-bn-in-2020-shows-data-120121700501_1.html#:~:text=Indian-\5.
2. Miclet, L., Bayoudh, S. and Delhay, A. (2014). "Analogical Dissimilarity: Definition Algorithms and Two Experiments in Machine Learning", *Journal of Artificial Intelligence Research*. vol. 32, no. 3., pp. 793–824.

3. Hosmer, D.W. and Lemeshow, S. (1989). *Applied Logistic Regression*. Wiley: New York.
4. Baobao, W., Jinsheng, M. and Minru, S. (April, 2008). "An Enhancement of K-Nearest Neighbor Algorithm Using Information Gain and Extension Relativity", *Proceedings of the International Conference on Condition Monitoring and Diagnosis*, pp. 1314–1317. IEEE, Beijing, China.
5. Bahety, A. (2006). *Extension and Evaluation of ID3 – Decision Tree Algorithm*. University of Maryland, College Park, MD.
6. Quinlan, J.R. (1986). "Induction of Decision Trees", *Machine Learning*, vol. 1, pp. 81–106.
7. Duan, H. and Liu, N. (December, 2013). "A Greedy Search Algorithm for Resolving the Lowermost C Threshold in SVM Classification", *2013 Ninth International Conference on Computational Intelligence and Security*, pp. 190–193. doi:10.1109/CIS.2013.47.
8. EMC Educational Services. (2015). *Data Science and Big Data Analytics: Discovering, Analyzing Visualizing and Presenting Data*. John Wiley & Sons: Indianapolis, IN.
9. Witten, I.H., Frank, E. and Hall, M.A. (2011). *Data Mining: Practical Machine Learning Tools and Techniques*, 3rd Edition. Morgan Kaufmann: Burlington, MA.
10. Simon, T. and Daphne, K. (2001). "Support Vector Machine Active Learning with Applications to Text Classification", *Journal of Machine Learning Research*. vol. 2, pp. 45–66.
11. Patil, D.V. and Bichkar, R.S. (2012). "Issues in Optimization of Decision Tree Learning", *International Journal of Applied Information Systems*. vol. 3, p. 22409-0868.

15 AI in Talent Management for Business Excellence

Subhajit Bhattacharya
Accenture

CONTENTS

15.1 Introduction 255
15.2 Proposed Model 256
15.3 Application Framework 258
15.4 Application Workflow 259
15.5 Key Challenges 259
15.6 Methodology/Process Followed 264
15.7 Critical Success Factor 265
15.8 Quantified Benefits to Business 265
References 266

15.1 INTRODUCTION

Human workforce plays an indispensable role in an organization, and in turn, talent management and workforce utilization become a crucial operation for the HR team. The overall talent management follows a holistic workflow starting from talent acquisition, background checks, onboarding and engagement, skill development, performance management, succession planning, workforce project planning, work allocation, workforce retention and smooth transition or exit processes.

In today's world when every organization is looking forward for its global presence and employees are working from different remote locations, it has become essential to ascertain optimized utilization of workforce. Every medium to large organization, today, is struggling to optimize the cost of operations while ensuring retention of the talented workforces and adopting intelligent solutions to automate delivery and replace human workforces with virtual workforces like robotic bots.

An obvious scenario is that as an organization starts expanding, workforce headcounts start growing, operations spread their arms across the geographies… the management of talent becomes challenging. Manual talent assessments and analytics are no more feasible nor advisable. Organization dynamics depends upon multiple factors that further contribute to an efficient workforce management.

In today's world, data and information together play a catalytic role to drive business smoothly and at scale. Every company at one hand is looking for cost cutting while at the same time trying its best to retain top talents and control the annual attrition rate at the lowest.

Today, leading organizations are rigorously working on the next-generation AI-led solutions for an end-to-end talent management which can perform all-round workforce orchestration and optimization at a real time. The AI advisory solution thus works in two different channels, first of all processing the real-time data through multilayered AI analytical engines for run time analytics and secondly cognitive assessments which work over the advanced machine learning modules to process large historic data and create correlation, perform deep learning, and prepare ambidextrous dashboards that enable to provide prolific data visualizations for decision-making (Eubank,,2018). This solution can predict if there is a need of additional recruitments, employee performance, and delivery forecasts and prescribe how workforces can be mapped to different projects to achieve the best resource utilization and optimization, future training needs based on the past and current performances, sentiment analytics to predict employee aspirations, commitment towards the organization, annual churn rates, etc., predictive pyramid refresh options for cost and delivery optimizations, intelligent 360° employee performance and appraisals, and many more. This AI solution and advisory tool is an out-of-the-box offering, which can be leveraged by most of the medium- and large-sized organizations to manage their human resources efficaciously (Russel, & Norvig, 2019).

This system is so intelligently designed that it can integrate and synchronize data from disparate platforms used for project management, service management, and workforce management. The system is highly scalable and agile in nature due to the fact that it can accommodate different types of a wide range of data and can be customized at much extent as per the organizational needs. The system is working on multitire security protocols over a hybrid SaaS model to give a glance of both online and offline data processing capabilities. Intelligent AI agents at the client end responsively get integrated with the client data and set a bridge between the client servers and the **IWOA Data Center** for data exchange and empower advanced AI-led analytics services.

The entire world is looking forward for such intelligent application which can help their workforce *mobilize, dynamize,* and *optimize* holistically, and thus, we call it **Intelligent Workforce Orchestrator and Advisor (IWOA)**. Renowned big ERP software manufacturing companies have shown their keen interest to elevate this product so that it can be well fitted with their people and project management modules.

In a summary, **IWOA** is the next-generation AI-enabled workforce and talent management software in the market to conquer with high-speed revolutionary analytical and advisory services on a hybrid SaaS platform to help the organizations grow with the talented people in the system at the most optimized level.

15.2 PROPOSED MODEL

IWOA is an augmented intelligence-led workforce optimization advisor for efficient project and service deliveries across industries. This is a unified multifaceted platform to analyze the performance, cost, productivity, and sentiment trend of workforces over the intelligent analytics platform to give a holistic view to the people manager to take needful decision and action proactively. This is a multitenant

AI in Talent Management

cloud-based solution (SaaS) working over secured and encrypted gateways to interact with the remote APIs residing at the clients' end to extract, replicate, integrate, and deploy workforce schedule and planning data into the IWOA application server to perform analytical assessments to generate analytical and interactive dashboards accessed over computer systems, tablets, and smartphones.

The key objective of IWOA is to help people managers to plan human resource management, perform skill gap analysis, and analyze people sentiments to boost morale through the advanced analytics platform.

IWOA is so enabled to get blended with an array of project, service, workforce, and schedule management tools and spreadsheets through custom APIs working over secured and autonomous channels ensuring encapsulation and abstraction among different sets of project data.

IWOA has an integrated identity management system to create and manage users, user types, and user groups and assign them adequate access rights and responsibilities.

IWOA also prepares log reports to track user activities and roll-back activities to ensure flawless and secured data transactions.

IWOA colab is a collaboration interface to exchange ideas and feedback over a larger platform, where the subscribers may interact with other associated subscribers, post blogs, and follow our latest updates and news.

IWOA Smart Interface is a next-generation product line wherein the AI analytics and communication platform are integrated to the smart devices for quick access and help our subscribers to receive vital alerts proactively.

IWOA is an AI-enabled solution that works on both the real-time and historic data to perform below major tasks:

- Data integration from hybrid data sources
- Intelligent workforce management advisor
- Employee sentiment analytics
- Employee performance trend analysis (time series assessment)
- Cross-functional skill performance assessment, skill gap assessment, and productivity analyzer
- Workforce skill and performance assessment (comparative assessment)
- Workforce operational cost optimization advisor
- Predictive and prescriptive advisory modules for workforce optimization, future project allocation, demand forecasting, and liquid agile workforce planning
- Advanced cognitive learning modules and ANN models for in-depth data assessments for intelligent prediction and prescription
- Advanced analytics:
  - Proactive alert liquid agile workforce fitment
  - Predictive performance assessment
  - Predictive training requirements reports
  - Predictive project delivery schedule assessments
  - Future workforce demand assessment
  - Scope for human resource re-alignment

User Type	Admin Roles		Data Management				Views
	IWOA Admin	Data Admin	Integration	Data Engineering	Analytics Business Logic		Dashboards & Reports
System Admin	Y	Y	Y	Y	Y		Y
Data Admin	N	Y	Y	Y	Y		Y
People Manager	N	N	Y	Y	Y		Y
Data Analyst	N	N	N	Y	Y		Y
MIS Executive	N	N	N	N	N		Y

FIGURE 15.1 IWOA user group and responsibilities.

In order to ensure security and user access levels, IWOA has different user groups and responsibilities (Figure 15.1).

- **System Admin**: This is a role-based group; the members will be having super user or admin rights to control active intelligence groups, users, data, data security controls, API configurations, business logic modules, dashboards and reports modules, and tracking activity audit trail reports.
- **Data Admin**: This is a role-based group; the members will be having elevated rights to control active intelligence data, data security controls, API configurations, business logic modules, dashboards, and report modules.
- **People Manager**: This is a role-based group; the members will be rights to control active intelligence data integration, custom query and procedure creation, business logic modules, dashboards, and report modules.
- **Data Analyst**: This is a role-based group, members will be rights to control active intelligence custom query and procedure creation, business logic modules, dashboards, and report modules including custom dashboard preparations.
- **MIS Executive**: This is a role-based group; the members will be having limited-view-only rights for dashboards and reports modules.

15.3 APPLICATION FRAMEWORK

IWOA is a hybrid SaaS application, wherein the application and the AI engine are hosted on the cloud server; however, IWOA intelligent agents are installed in the local workstations which further interact with the databases at the user end through the app/database gateways and the data integration APIs. In order to transfer and synchronize the data to and from the IWOA server to the user database, it uses data replication and integration services.

IWOA can integrate with a large array of applications like spreadsheets, team collaboration software, HR systems, project management software tool, workforce management tool, ERP applications, and databases.

At IWOA cloud, AI-enabled data mining modules perform the ETL process to extract the data from the data lakes and segregate them into different data clusters; following this, the processed clustered data are pushed into information bases post multiple levels of data correlation and data symmetric analytics, which help to prepare multifaceted reports and dashboards. Information from the information bases is further processed by intelligent data processing units augmented with AI engines to perform AI analytical services and AI advisory (Soundararajan, & Singh, 2016).

In order to access the IWOA application, we have various types of users; however, every user is mapped with a user role, and every role may have standard and custom responsibilities (Figure 15.2).

The overall high-level architecture can be well depicted in Figure 15.3.

Besides the end-to-end data encryption, user level security has been provisioned. Therefore, the user persona can be visualized in Figure 15.4.

15.4 APPLICATION WORKFLOW

IWOA has various modules to enable users to setup portfolio, programs, and projects along with user setup, user group setup, API configuration, data source setup, data model preparation on the fly, business logic setup, data integration and synchronization configuration, query builder, KPI and metrics setup, IWOA trigger setup, etc.

It is important to understand, how the entire application works. The high-level application workflow is shown in Figure 15.5.

15.5 KEY CHALLENGES

The problem related to workforce optimization and task orchestration was realized post having discussions with the leaders of many companies clustered either under IT or non-IT industries. Their problems were relevant and realistic as human workforces are the fuel to the companies, and an intelligent governance system can ensure a business continuity by virtue of continuous quality-led timely deliveries which not only delight the clients or customers but should also induce them to be innovative as well.

In order to combat the problem, IWOA, a digital platform has been introduced that aims to improve workforce utilization and workforce engagement in real time. Further to this, below excerpts from various well-known journals did talk about the immense opportunities of AI to optimize workforce within an organization (Marr, 2018).

a. **Enterprise Embracing Big Data and AI**: According to HBR (2019), there is a significant increase in investment in big data and AI initiatives. Among 5,000 executives, 92% of survey respondents reported that the pace of their big data and AI investments is accelerating; 88% report a greater urgency to invest in big data and AI; and 75% cite a fear of disruption as a motivating factor for big data/AI investment. In addition, 55% of companies reported that their investments in big data and AI now exceed $50 MM, up from 40% just last year. Further, companies are building organizations to manage their big data/AI initiatives, with a rise in the appointment of chief data officers from 12% in 2012 to 68% of organizations having created and staffed this role in the past 7 years.

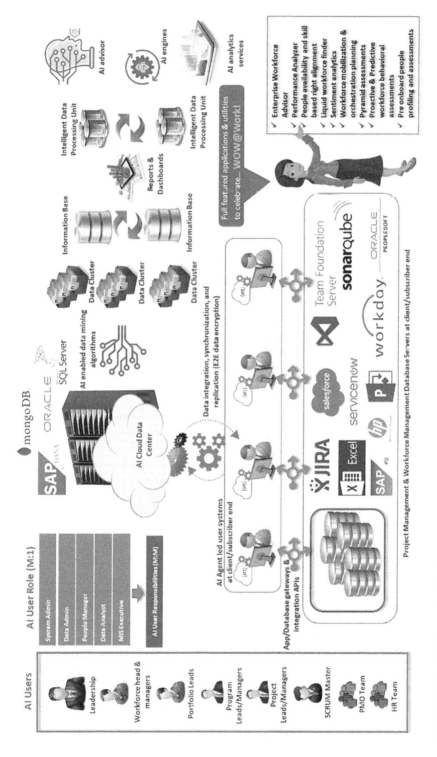

FIGURE 15.2 IWOA application framework.

AI in Talent Management

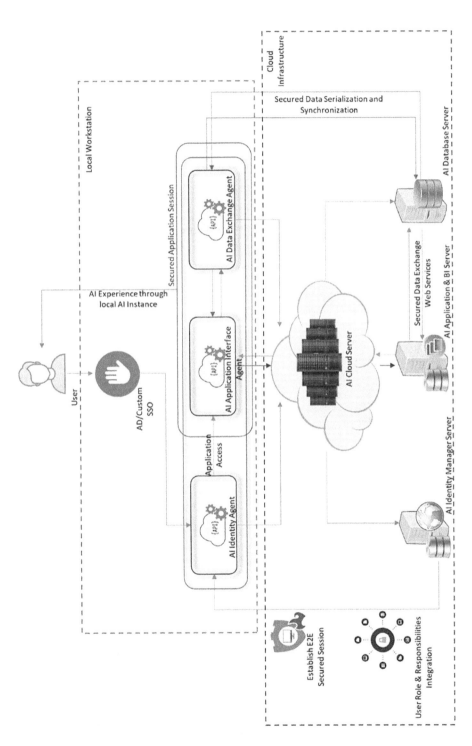

FIGURE 15.3 IWOA application architecture (high level).

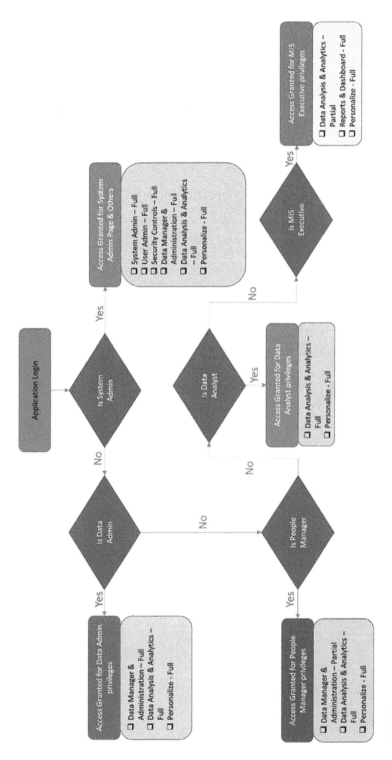

FIGURE 15.4 IWOA user type and security.

AI in Talent Management

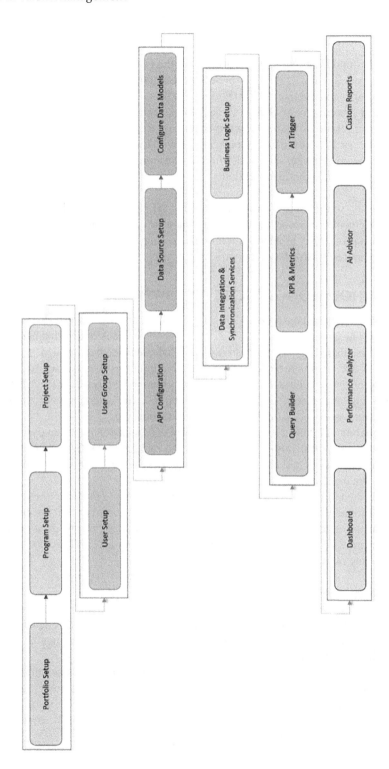

FIGURE 15.5 IWOA application workflow.

b. **Business Intelligence**: Improving revenues using BI is now the most popular objective enterprises are pursuing in 2019, while helping better decision-making is the upmost function for the IT industry. Enterprises with the largest BI budgets this year are investing more heavily into dashboards, reporting, and data integration (Forbes, 2019).
c. **Project Management Office**: Project Management Institution (2018) conducted a research regarding top three factors that threaten company project success. First, insufficient engagement with executive sponsors; second, project scope change; and third, immature resource capability and low efficiency project resource utilization.

The challenges, so far, were seen to be a missing AI platform for end-to-end workforce management and information transparency to enable complex self-learning systems to act proactively as an advisor module and may produce an array of analytical reports and dashboards based on various scenarios.

15.6 METHODOLOGY/PROCESS FOLLOWED

IWOA is an out-of-the-box AI-enabled solution, which is developed over a hybrid SaaS model. The application uses intelligent agents at the client system that performs multiple tasks, including user authorization, establishes secured session between IWOA server and client system, enables data collection from client server, regulates data replication and synchronization from client system to IWOA server for real-time data processing, and enables the client IWOA app to act in a serverless mode, thus enabling offline data processing, analytics, visualization, etc.

IWOA analytics engine at the client system works as an autonomous analytics app which ensures data serialization and real-time data cache from IWOA Lite Database residing at the client end and induces various analytical services for report and dashboard, performance analyzer, and AI advisor.

Data collected at IWOA servers are further processed through AI engines to perform data anomaly detection, cleansing and auto-correction, data mirroring, correlation, clustering, cognitive assessment for data knowledge pattern creation, and machine learning knowledge repository update (Edwards & Edwards, 2019).

The ultimate objectives of IWOA are as follows:

- Intelligent workforce management advisor
- Workforce utilization
- Bench optimization for proactive resource utilization and project skill alignment
- Predictive project risk assessment (workforce dependency assessments)
- Employee sentiment analytics
- Employee performance trend analysis (time series and behavioral assessments)
- Cross-functional skill performance assessment, skill gap assessment, and productivity analyzer
- Inter-workforce skill and performance assessment (comparative assessments for appraisals and rewards)
- Workforce operational cost optimization advisor

AI in Talent Management

- Predictive and prescriptive advisory modules for workforce demand forecasting and liquid agile workforce planning
- Advanced cognitive learning modules and artificial neural networking models together blended for in-depth data assessments inducing real-time intelligent advisories for ensuring workforce availability and alignments for timely project deliveries, project operational cost optimization, upskill requirements, hiring requirements, talent best-fit analysis, future workforce demand assessments, etc.

15.7 CRITICAL SUCCESS FACTOR

To ensure the best usage of IWOA resulting in the best-in-class results, below, yet not limited to, checkpoints should be strictly considered:

- Organization details
- Organization operation details
- Structured details of employees
- Employee roaster
- Employee cost to company details
- Portfolio and project details
- Project knowledge books
- Consolidated project and operation delivery plans and employee mapping
- Project pipeline details and human resource demands
- Yearly project revenues
- Current project backlogs
- Key operations and process KPIs
- Employee KRAs
- Delivery OLAs and SLAs
- Employee skillsets
- Bench register
- Past employee annual performance
- Current and past employee details along with the project skills.

Being a perfect hyper AI-led solution, IWOA would be at its best while we consider both the past and current data for retrospective and predictive data analysis. Intelligent data mining algorithms at the server end blend these stacks of data as training datasets, thus making the AI and analytics engines work more robust and tend to be accurate while performing predictions and advisories on different scenarios and queries.

15.8 QUANTIFIED BENEFITS TO BUSINESS

Some of the foreseen key benefits of IWOA applications are as follows:

- Smart talent acquisition and retention planning
- Predictive training and upskill needs

- Multifaceted employee performance assessments and consolidated score cards at real time
- Predictive project risk assessments due to team dynamics and dependencies
- Workforce behavioral and sentiment analysis
- Optimized workforce allocation and project demand fulfillment
- Virtual workforce advisor for the scenario-based solution
- Workforce task orchestration for real-time workforce optimization

Therefore, we can very well conclude that IWOA is a perfect hybrid SaaS-enabled application wherein organizational data get converged to produce both qualitative and quantitative outcomes for the organizational talent retention, workforce optimization, revenue generation, customer satisfaction, and most importantly organization expansion across different geographies as irrespective of the human resources working from any of the remote locations, the IWOA virtual agents can very well track and prescribe the best fit of the talents to ensure a mutual growth of both the organization and people.

REFERENCES

Martin Edwards, Kirsten Edwards. 2019. *Predictive HR Analytics: Mastering the HR Metric.* Kogan Page, New York.

Ben Eubanks. 2018. *Artificial Intelligence for HR: Use AI to Support and Develop a Successful Workforce.* Kogan Page: London.

Bernard Marr. 2018. *Data-Driven HR: How to Use Analytics and Metrics to Drive Performance.* Kogan Page, New York.

Stuart J. Russel, Peter Norvig. 2019. *Artificial Intelligence – A Modern Approach.* Pearson, Berkeley, CA.

Ramesh Soundararajan, Kuldeep Singh. 2016. *Winning on HR Analytics: Leveraging Data for Competitive Advantage.* Sage, London.

https://www.accenture.com/in-en/services/applied-intelligence/mywizard-intelligent-automation-platform.

https://hrcurator.com/2020/07/08/how-ibm-incorporates-artificial-intelligence-into-strategic-workforce-planning/.

Index

Note: **Bold** page numbers refer to tables and *italics* page numbers refer to figures

additive manufacturing 99, **108**
advance manufacturing **8, 10,** *12*
artificial intelligence (AI) 52, 54, 62, 67, 75, 255
    machine learning **108**, 109
    platforms 62, 79, 81
    and retail 77
augmented reality 49, 86, **89**, 90
autonomous vehicles 115, 123, 124

behavioral economics 208
big data 175
    and AI 259
    analytics 198, *198*
    lifecycle 179, *179*
bitcoin 207, 208
blockchain 141, 151, 154, *155*
    based IoT network 157, *155, 158*
    and digital payments 196
    technology 197, *197*, 198, 211
boxplot 241, *243*
business intelligence 264

chatbots 49, 52, 54, **56,** 81, *82*
cloud 143
cluster analysis 9, *9*
competitive advantage 103, 104
component of IoT 142, *143*
content
    analytics 2
    automation 84, 88, 89, *90*
continuous improvement 98, 103, 110
Covid 19 237
cryptocurrency 205
customer
    data 180, 186
    experience 47, 48
    perception 62
    satisfaction 83

data
    flow *41*
    mining **131**
    security risk 194
    synchronization network 27, 31
decision tree 240, 251
digital
    payments 191
    transformation 47
digitization 98, 103

economy 163
efficient consumer response 27, 28
Elastic Net regression 226, 232, **233**
emerging technologies 196
evolution of big data 178

face recognition 48, 52
facilitators of digital payments 195
fast moving consumer goods 31
financial services 175, 181
foreign exchange 164
Forex 166, 167
fourth industrial revolution 17

game theory 163, 165
global ECR scorecard 31, *32*
grey DEMATEL 134

heatmap 230, *230*

indoor logistics 121
industrial Internet of things (IIoT) 127, **130**
industry 4.0 technologies **108**
inhibitors of digital payments 193
Internet of things (IoT) **75,** 142, 159, 180
    architecture 146
    communication model 151, *151*
    security 141, 148, *149*
    vulnerabilities 152
IWOA 256, *261, 262, 263*

K-Nearest Neighbors 240

Lasso regression 226, 232, **233**
lean
    and industry 4.0 98, 100, 104
    operations 97
    principles *101*, 110
    tools 102, 103, 104
linear regression 226, 233
logistic regression 239

machine learning 163, 165, 227, 238, 239
mining tweets *5*

Naïve Bayes 241
network
    analytics 6, 21
    flows 31, *35*
    graph 15, *15, 16*
new age technologies 47, 48, 49

online retail 62, 63, 90
outdoor logistics 122

personalization 52, 55, 57, **89**, *90*
prediction 166, 168
predictive training 265
product
    category management 30, 36, *37*
    suggestions 52
project management office 264
psychological risk 194
Python 226

retail trade 36, 38, *38*
ridge regression 226, 228, **233**
risks 115, 124

scatter plot *229*
sentiment analysis 12, 20
smart talent acquisition 265

social media 2
social risk 194
SVM 240, 241

talent management 255
technology adoption 54, 56, 57
time risk 194
Twitter Analytics 1, 2

universal product coding 27

virtual reality 103, 110
visual search 90, *90*, 91
voice assistance 74, **89**
voice search 89, *90*, 91

wealth management 212
WEKA 224, 225
word cloud 242, *243, 244*